WHAT GOES UP...
GRAVITY AND SCIENTIFIC METHOD

The concept of gravity provides a natural phenomenon that is simultaneously obvious and obscure; we all know what it is, but rarely question why it is. The simple observation that "what goes up must come down" contrasts starkly with our current scientific explanation of gravity, which involves challenging and sometimes counterintuitive concepts. With such extremes between the plain and the perplexing, gravity forces a sharp focus on scientific method.

Following the history of gravity from Aristotle to Einstein, this clear account highlights the logic of scientific method for non-specialists. Successive theories of gravity and the evidence for each are presented clearly and rationally, focusing on the fundamental ideas behind them. Using only high-school level algebra and geometry, the author emphasizes what the equations mean rather than how they are derived, making this accessible for all those curious about gravity and how science really works.

PETER KOSSO is a philosopher of science. He taught physics at Montana State University, and taught philosophy first at Northwestern University, and then at Northern Arizona University. He is the author of *Reading the Book of Nature*, *Appearance and Reality*, and *Knowing the Past*, as well as numerous articles on relativity, quantum mechanics, astronomy, and scientific method.

WHAT GOES UP . . . GRAVITY AND SCIENTIFIC METHOD

PETER KOSSO

Northern Arizona University (Retired)

CAMBRIDGE
UNIVERSITY PRESS

CAMBRIDGE
UNIVERSITY PRESS

University Printing House, Cambridge CB2 8BS, United Kingdom

One Liberty Plaza, 20th Floor, New York, NY 10006, USA

477 Williamstown Road, Port Melbourne, VIC 3207, Australia

4843/24, 2nd Floor, Ansari Road, Daryaganj, Delhi - 110002, India

79 Anson Road, #06-04/06, Singapore 079906

Cambridge University Press is part of the University of Cambridge.

It furthers the University's mission by disseminating knowledge in the pursuit of
education, learning and research at the highest international levels of excellence.

www.cambridge.org
Information on this title: www.cambridge.org/9781107129856

First published 2017

Printed in the United Kingdom by Clays, St Ives plc

A catalogue record for this publication is available from the British Library

Library of Congress Cataloging-in-Publication data
Names: Kosso, Peter, author.
Title: What goes up ... gravity and scientific method / Peter Kosso, Northern Arizona University (Retired).
Description: Cambridge, United Kingdom ; New York, NY : Cambridge University Press, 2017. |
Includes bibliographical references.
Identifiers: LCCN 2016032822| ISBN 9781107129856 (hardback) | ISBN 1107129850 (hardback)
Subjects: LCSH: Gravity. | General relativity (Physics)
Classification: LCC QC178 .K87 2017 | DDC 531/.14 – dc23
LC record available at https://lccn.loc.gov/2016032822

ISBN 978-1-107-12985-6 Hardback

Contents

Preface

Gravity is arguably the most obvious and most obscure of natural phenomena. What goes up must come down. The current scientific explanation of why this happens, the general theory of relativity, invokes challenging and counterintuitive concepts like curved geodesics in a four-dimensional spacetime. With such extremes between the apparent and the arcane, gravity forces a sharp focus on scientific method. The relation between observation and theory is the heartbeat of science, and the pulse is nowhere stronger than in comparing what we observe and what we theorize about gravity. This is the motivation for a book about both the science of gravity and the scientific method.

If gravity is such a simple and obvious phenomenon, why complicate things with obscure theory? If gravity is so undeniably real and so easily observed, as in, go jump off a cliff and then tell me the law of gravity is a social construct, why does the basic description, the theoretical account of gravity, change? How could the early scientists have missed something so obvious as a force between two massive objects? But by the light of general relativity, it's not a force after all. So, maybe it's not so obvious.

What Goes Up ... Gravity and Scientific Method will clarify the theories of gravity from Aristotle to Einstein. Aristotle's explanation was that a stone falls because it seeks its natural place at the center of the universe. According to Newton, a stone falls because of an instantaneous force from the massive Earth. Einstein, and most physicists now, say a stone falls because, with no forces acting, it follows the curved geodesic in spacetime. Differences and similarities between the theories will be highlighted, and there will be some surprises. For example, Aristotle's idea that the trajectory of the falling stone is guided by a point in space, the center of the universe, is not so different from Einstein's claim that the trajectory of the fall is guided by a line in spacetime, the geodesic.

Equally important to understanding these scientific results will be understanding the scientific methods. By showing how the theories were derived and tested, we will clarify what makes science scientific. The focus will be on the relation between theory and evidence, and links among theoretical ideas and principles. Again, methods will be compared, and again there will be surprises. Aristotle is often criticized for deriving his scientific ideas from principles rather than evidence, and for failing to test his theories. Einstein did very much the same thing by starting with the Principle of Relativity and the Principle of Equivalence, and he is famously dismissive of the importance of the 1919 solar-eclipse evidence, one of the so-called classic tests of the general theory of relativity.

Once a few main questions and concepts are in place, the development of the theories of gravity will be in historical order, from ancient Greece to the present, with one exception. The survey starts with Newton. The first two chapters raise the relevant questions about scientific method and introduce the helpful scientific concepts to describe gravity, but then the third chapter begins the study of gravity with a review of what you would find in any introductory physics textbook, that is, the Newtonian theory. It will be out of historical context and with no justification or proof, not a word about scientific method. From there, we go back to the historical beginning and follow the development from Aristotle to Newton, this time with both context and motivation, to Einstein and current developments. The reason for doing the basic Newtonian theory first is to start with the familiar. When most people think of gravity they think of Newton, and maybe the acceleration of gravity at 9.8 m/s^2 and the basic force law $F = GMm/r^2$. But they generally don't ask where the ideas came from or how they differ from what came before and after. The force of gravity is taken for granted. We start by making the familiar precise, so we have a sense of our perspective and no longer take it for granted. The first look at Newtonian gravity will give us some conceptual context, a way to compare what came before and after. It's wise to do history with your current ideas out in the open. That way we'll see if our description of the past is being influenced by our understanding of the present.

There are no prerequisites for reading the book, neither scientific nor mathematical. Technical terms will be minimal and always explicitly clarified. Any important concepts, whether scientific or philosophical, will be printed in **bold** when they are first described in the text, and then defined in the Glossary. There is a little math, but it is never more challenging than algebra and geometry. The math will be limited and in all cases avoidable by readers who find it off-putting. It will be there for those who enjoy it and for everyone to at least see what work in gravity looks like to those doing the work. We won't derive any of the equations, or even use them to do calculations. The emphasis will be on understanding the physical implications of the components of an equation. For example, the r^2 in the equation in the previous paragraph, the $1/r^2$ dependence in the law of gravity, is important. It's not just that the force gets weaker as you get further away; it decreases as the *square* of the distance. We'll find out why the $1/r^2$ is there, and what it leads to for phenomena like planetary orbits and escape velocity. By the time we get to the general theory of relativity, this sort of careful accounting of what work is done by each piece of the equation will clarify what it means to say that gravity is the curvature of spacetime.

This book started as a class in the Northern Arizona University Honors Program. I have the Honors Program to thank for the inspired proposal of a class that requires both substantive science and a study of how science works, and for allowing me the opportunity to teach the class with gravity as the focus. The students in the class were a great help in pointing out where things were confusing and in many cases clearing up the confusion. In that formative way they helped write *What Goes Up ... Gravity and Scientific Method*. Several of my colleagues at Northern Arizona University helped along the way with conversations about relativity, astronomy, and the history of science. Thanks to David Sherry in the philosophy department, to Ed Anderson and Gary Bowman in physics, and to Andrea Holmen in both.

I should also thank a few people with whom I had the pleasure to work at Cambridge University Press. Thanks to Vince Higgs, to whom I pitched the proposal and who had the kindness to first improve my pitching and then take on the project. And to Philippa Cole, whose correspondence I always looked forward to, for both the cheerful encouragement and detailed help.

Some of the material in the book is taken from my previously published work. Parts of Chapters 4 and 5 come from "Void points, rosettes, and a brief history of planetary astronomy," *Physics in Perspective*, **15** (2013), 373–390, used with the kind permission from Springer Science+Business Media. The first half of Chapter 8 is from "The discovery of Neptune," *School Science Review*, **90** (September 2008), 53–58, by permission from the Association for Science Education. Chapter 13 has pieces from "Detecting extrasolar planets," *Studies in History and Philosophy of Science*, **37** (2006), 224–236, and Chapter 14 comes in part from "Evidence of dark matter and the interpretive role of general relativity," *Studies in History and Philosophy of Modern Physics*, **44** (2013), 143–147. Both are by permission from Elsevier.

1

Introduction: What to Expect from a Science of Gravity

Gravity dominates our lives and attracts our attention like no other force of nature. First thing in the morning, just getting up and out of bed, lifting your head from the pillow and dropping your feet to the floor, it's gravity that works with you and works against you, and you know it. You will spend your day opposing and cooperating with gravity, lifting the coffee pot, pouring the coffee, and so on.

There is no up or down without gravity. This most basic direction is defined, not by some cosmic or even planetary coordinate system, but by the force of attraction between the Earth and things. The force is generally directed toward the center of the spherical planet; that's down. The other direction is up. A builder determines that a wall is vertical by using a plumb-line, a mass hanging free in the field of gravity. And for all of us, getting up in the morning amounts to changing our orientation in the gravitational field from horizontal to vertical, from lying down to standing up.

Gravity is the force we all notice and the one we can all identify explicitly. This is despite the fact that physicists describe gravity as the weakest of the four fundamental forces in nature. The electromagnetic force is responsible for holding the atoms and molecules together in everything we touch, but this is largely overlooked as we go about our days. Two versions of nuclear force are responsible for stabilizing the core of matter, the atomic nucleus, or in some cases for destabilizing it, perhaps the start of the process of heating your coffee if your electricity comes from a nuclear power plant. But the nuclear glue holds at such short range as to go undetected unless you know what to look for. It's only gravity that we all experience and we all know.

The force of gravity is always attractive. It is a force pulling together any two things that have mass. Any amount of mass will cause the attraction, but the more mass the stronger the force. That's why a brick is heavier than a balloon, heavier and harder to hold or move in opposition to the force of gravity. The force also depends on the distance between the objects, the greater the distance the weaker the force. The attraction gets weaker and weaker as the objects get further apart, but it never disappears altogether. The force never goes to zero. And it never pushes things away. That's because objects always have a positive amount of mass. There is no such thing as negative mass. It's not like electric charge that comes in both positive and negative values.

The electromagnetic force works between any two things that have electric charge. The more charge, the stronger the force. And, like gravity, the force depends on the separation between the objects. The further they are apart, the weaker the force. But unlike gravity, electromagnetism can be both attractive and repulsive; it can pull things together and push them apart. Opposite charges, one positive the other negative, attract. Like charges, both positive or both negative, repel. This means that a composite object that has an equal amount of positive and negative charge will push and pull in equal amounts and consequently experience and exert no electromagnetic force at all. This balance, the result of charge neutrality, can never happen with gravity. There being no negative mass, there is no possibility of a mass-neutral object. You can't neutralize gravity as you can the electromagnetic force. This is why it is ever-present in our experiences.

Atoms are held together by electromagnetic forces. The nucleus is positively charged and the electrons are negative, hence there is a force of attraction between them. But there is generally a balance, an equal number of positive protons in the nucleus and negative electrons in orbit, so the whole atom has no electric charge. Thus, there is little electromagnetic force beyond the internal bonding of the atom itself. There is a little, and it's what holds atoms together as molecules and ultimately the solid objects we encounter in experience. It results from charges within the atoms being unevenly distributed, a little negative on one side and a little positive on the other, or the outright loss or sharing of an electron. The result is that the stuff we encounter is electrically neutral. We do not experience the electromagnetic force at work on the very small structural level.

If you shuffle your feet across the floor, you will be reminded of the electromagnetic force. Scraping electrons from the carpet will give your body a small negative charge that will quickly flow to a conductor like a doorknob. You get a shock. So, sometimes electricity is as obvious as gravity, but these times are rare. The electromagnetic force usually hides, since charges are balanced and objects are neutral. Gravity can't hide.

Deeper inside the atom, in the nucleus, the pieces are held together by what is plainly called the nuclear force. Electromagnetic forces won't do, since there are only the positively charged protons and uncharged neutrons to work with. Whatever holds these together has to be a very strong force to overcome the electrical pushing apart of the protons, and at such close quarters. The nuclear force is strong, but it is confined to particles like protons and neutrons. It has no effect beyond the nucleus itself. Because of this, it is hidden from our daily experiences.

That leaves gravity as the only basic force to dominate our awareness of the natural world. There is nothing more basic than knowing up from down, nothing more basic than gravity.

Gravity governs our personal and mundane activities; it also controls the grandest events of the cosmos. Planets and moons, like spaceships and stars, have no electric charge. In this way they are like atoms. Consequently, the only force that affects them is gravity. The force of gravity is of infinite extent. It gets weaker at distance but it never lets go entirely. It's a force even at astronomical distances. Gravity holds the Moon in orbit around the Earth and the Earth in orbit around the Sun. It is responsible for the dynamics of galaxies and galactic clusters and clouds of intergalactic gas and dust.

Gravity reveals our cosmic past and determines the future. The universe is expanding, and the rate of expansion is controlled by gravity. If we ask when the expansion began, the answer is calculated with the knowledge of the behavior of gravity. And if we ask how long the expansion will continue, and what will happen then, the calculation requires an understanding of gravity, and pretty much gravity alone. There are unknowns. Perhaps the universal force of gravity is strong enough, and the current rate of expansion slow enough, that the expansion will slow down and stop and reverse. The universe will collapse and reassemble to the condensed condition of the big bang. Or maybe the expansion is too great, or even speeding up, and will continue forever. There was a beginning of time, but there will be no end. These are profound cosmological questions with deep metaphysical implications, and only an understanding of gravity can deliver the answers. It's gravity that controls the situation.

So gravity is universal and ubiquitous. It's with us through the day, sometimes for us and sometimes against, and we know it. What goes up must come down. So what's left to discover? What do we expect or need from a science of gravity, when we are personally so familiar with the force? Who needs a scientist to tell them which way is up?

We can address this issue with an eye toward the more general question of what to expect from a science of anything. Are there some kinds of things and events that can be studied scientifically and some that cannot? What sorts of results can we expect of a scientific study? Science will describe aspects of nature, but will it also explain them? And what exactly are the standards and methods required for the study to be genuinely scientific? What makes science scientific?

Look for two things from a science of gravity, a more precise and thorough description of gravitational phenomena, and (maybe) an explanation of those phenomena. The description will provide more detail than we notice and note in our casual encounter with gravity, and the data will be more carefully gathered. The description may even go beyond what can be observed, to report some underlying mechanism. And this may lead to the explanation. We may be observant and pretty clear on the effects of gravity, but the cause is less apparent.

The distinction between description and explanation, and the proper role of each in the process of science, is itself not so clear. We'll work on this as the science of gravity is developed in subsequent chapters. Rather than prescribing a methodological form for science to follow, we'll let the basic structure develop under the influence of watching the science at work. This is similar to what happens in science itself; you don't force the data into a rigid theoretical form. Rather, you let the theory take shape under the influence of the data. But basic, revisable theoretical ideas are needed in the beginning if only to direct the gathering of evidence, to know what to look for and when it has been reliably observed. Similarly, we should have some preliminary ideas about the roles of description and explanation in science before the science starts.

The centerpiece of Isaac Newton's science of gravity is the law of universal gravitation. Any two objects, anywhere in the universe, attract each other with a force proportional to the product of their masses and inversely proportional to the square of the distance between them. What is this other than a summary of a regularity in nature, a purely descriptive summary? There is no suggestion of *why* there is this force of attraction, or why it is linked to mass rather than, say, color or smell. There are only the descriptive details, universal and precise. One textbook on gravity makes this a general point about science: "As with any physical law, there is no reason 'why' the world had to be this way..." (Bernard Schutz, 2003, p. 3). The suggestion is that laws, a principal currency of a science like physics, describe but they do not explain.

Perhaps answering Why? questions is beyond the legitimate purview of science. Asking for the reason some aspect of nature is the way it is may cross a line from the natural to the unnatural, a first step toward matters occult and metaphysical. But surely some scientific accomplishments amount to explanations. Why is the sky blue? You wouldn't just say, Well that's just the way it is. There's a legitimate scientific explanation to give. The sky is blue because molecules in the air scatter the short-wavelength blue light more effectively than the long-wavelength red. When you look at the sky away from the Sun, you see the scattered light, the blue. The details of description, in this case including the unseen components, the molecules that scatter the light and the electromagnetic wave that is the light, reveal the mechanism that causes the phenomenon. And there are lots of examples of scientific explanations like this. Arguably, the primary task of medical science is to identify the causes of particular diseases, that is, to explain symptoms, explain why a person is sick and suffering in some particular way.

The distinction between description and explanation is not so clear that we want to count one as legitimate science and the other as suspect or beyond the limits of science. Let's wait and see what happens in the science of gravity, what gets described and how, and what, if anything, gets explained.

What gets described, and how? To describe something *scientifically* will require a standardized language of description. Informally, that is, *not* scientifically, we can get by with colloquial terms like heavy or light, fast or slow, up or down. But these are vague and subject to changes as used by one person or another, or even from one event to another in the life of one person. A prerequisite of a scientific description of gravity is a stable, precise, and intersubjective vocabulary. It is natural to involve numbers and mathematics in this descriptive task, since numbers are stable, precise, and intersubjective like nothing else. Whether the car is going *fast* may be a matter of interpretation or context, but its going 100 mph is a matter of fact.

One of the many virtues of using clear and rigid terms, quantifiable when possible, is that it facilitates testing scientific claims. Vague claims are hard to test. More to the point, they are hard to disprove. If one person's fast is another's slow, the prediction that a stone's fall will be fast will turn out to be true for one but false for another. But with clear terms like acceleration and velocity, and precise parameters like meters and seconds, results of the testing will much more likely be agreed to by all parties.

Shared terms with unchanging meanings also allow repeatability in science. This, like testability, is often cited as a basic scientific requirement. Results in my lab or observatory have to be repeatable in yours. This is not just a guard against fraud, but against systematic error in my work. But I have to be able to tell you how the experiment or observation was done, and this must be in clear and exact language that you and I understand in exactly the same way.

Precise and repeatable terminology also helps in practical matters dealing with the forces of nature. To avoid the dangers of gravity or to make use of its powers, we need to anticipate effects and share information. Galileo knew this. Many of his contributions to the science of gravity were motivated by his employment to optimize the range and accuracy of cannons. Plotting trajectories in mathematical terms, that is, with precision and uniformity, revealed the parabolic path of a projectile like a cannon ball. This in turn tells the soldier the angle to tip the gun to send the ball the needed distance.

The language used to describe what is observed in science is more precise than in day-to-day experience; so is the act of observation itself. None of us, I hope, is so naïve as to accept the Wile E. Coyote trajectory when running off a cliff. It's not straight out, pause (time to lament), then straight down. It's an arc of some form, curving out and down at the same time. But few of us are so focused in tracking a ball that rolls off a table as to note the parabolic shape of the arc. This precision aided Galileo and other ballistic scientists; it is also an essential start to doing rocket science.

Scientific observation is thus more systematic and detailed than our informal account of the natural world; it is also more extensive. There is more to see than meets the eye. Microscopes, telescopes, and other tools of observation enrich the evidence and extend our empirical reach. This scientific commonplace is not so common in the science of gravity. There is no graviscope. Astrophysicist Evelyn Gates describes what she calls "Einstein's Telescope," but this is not a device built to magnify or probe the otherwise invisible details of gravitation. It's a technique that takes advantage of the gravitational effect known as lensing, bending light from distance stars or galaxies, to find focused images of the light sources. It's more about finding the lens, the gravitating mass that's doing the bending, than about magnifying the specimen, the star or galaxy. But it is a way of finding large masses in the universe that would otherwise be undetected.

Extending the scientific description of nature even further we get to things that simply cannot be observed, even in an indirect way by using instruments like microscopes and particle detectors. The physicist's description of nature at its most fundamental level is composed of quarks, fields, spacetime, virtual particles, and more. The world is, in a sense, nothing like it appears to be. The solid table is in fact mostly empty space. And even the bits that are not empty are not really bits. The atoms, made of electrons, protons, and neutrons, and the protons and neutrons made of quarks, are not tiny specks of solid matter so much as they are diffuse fields of probability. The scientific description, in other words, is meant to get us to the underlying reality that makes up the easily accessible appearance.

Back to what we can observe, the scientific description of the evidence is further refined by explicitly separating individual components from a composite experience. We see things fall. Science sees the action of two things, gravity and air resistance. We see a kicked ball roll to a stop. Science sees inertia and friction at odds. A feather falls more slowly than a stone, and we might leave it at that. But the science of gravity clarifies the need to distinguish the action of gravity from that of air resistance. Isolating the causes allows us to understand each. If there was no air, if it was only the force of gravity at work, the feather and stone would fall at the same rate. This is a property of things that is so important that is gets a proper name, the Principle of Equivalence. It will play a fundamental role throughout our study of gravity.

On our own we see and can roughly describe what happens in the natural world around us. From science we expect more, that is, a description of what *does* happen and of what *would* happen in different circumstances. Science should be able to generalize and abstract observed information and project onto unobserved situations. Knowing about gravity and air resistance, it is possible to say what would happen to the feather and the stone in a vacuum. This sort of expansive reasoning requires a fundamental understanding of the forces in play in order to separate the essential properties from the circumstantial. The law of gravity has a role for the masses of the objects, and shows that the gravitational acceleration must be independent of mass. Natural laws, not unlike the laws of a society, include a level of necessity. It's not that all cars do stop at stop signs. All cars *must* stop at stop signs. What goes up *must* come down. It's the necessity and the complete generality that makes this a law. And in the case of traffic, it's simply being a car that counts. The color of the car, the make, the income of the driver, and so on, are irrelevant. A law picks out the essential properties of application and ignores the inessential. And this allows the extension to unrealized cases. If a car, any car, were to encounter a stop sign it would be required to stop. Similarly with the law of gravity. What would happen if we dropped a feather and a stone on the Moon? Knowing the mass of the Moon and the airless conditions, and applying the law, the two would fall at the same, precisely specified, rate.

Rocket science depends on the ability to describe and predict the action of gravity under different conditions. The Moon has a different mass than the Earth, so its gravitational effect will be different. A rocket will be influenced by varying forces at varying distances from the Earth. So, mass and distance are relevant to what *would* happen to the rocket. But rocket scientists don't care about the color or temperature or even the elemental composition of the objects of their calculations. Again, a basic understanding of the law of gravity tells them which questions are important and which are not worth asking. What would happen if the satellite was twice as far away? That's a good question. What would happen if the satellite was blue? That's a bad question, a waste of time, an experiment not worth funding and a parameter not worth controlling. In this way, the science of something refines the study and the description by directing the questions to be asked. It tells us what can be safely ignored.

Scientific expertise directing the questioning is both a good thing and a bad thing. Aristotle studied gravity, and his broad scientific understanding focused the questioning on the elemental composition of objects. In Aristotelian science, earth, air, fire, and water, or some combination of these four elements, make up everything around us. The element earth naturally seeks the center of the universe, and hence falls toward the center of the Earth. Fire naturally moves away from the center, and so hot and tenuous things rise. From this Aristotelians developed a coherent account of the observed motions of things. But we would say the Aristotelian account is false and that he focused on the wrong properties of things. The danger in allowing a science to direct the questioning is that the answers might too easily conform to expectation.

Here is the first indication that doing science, even if the results are purely descriptive, is not a matter of observation alone. Good observation, *scientific* observation, requires insight and expertise. It requires some theoretical understanding of what to look for, how to look, and what it means when you find it, or don't. Scientific method will not be a purely bottom-up process in which theoretical conclusions are based firmly on independent observations. There will have to be an ongoing give-and-take between theory and observation, neither strictly prior or independent or more authoritative that the other. The reciprocal relation will be key, and it's these details we need to watch as the science of gravity unfolds.

What can be observed regarding gravity? Is it accurate to say that the phenomenon of gravity itself is something we can observe and in fact do observe every day? Or is gravity in some sense one of nature's hidden properties that we can only understand indirectly, as we can know about quarks and black holes only by their effects on other things we can experience? To say that gravity is a ubiquitous aspect of our lives is not to imply that the force itself is directly observable.

If gravity were plainly observable, there should be no doubts or controversies about it. There would be nothing to discover, if the mechanism were there to see or feel directly. We would have known about it all along, or at least since humans started keeping track of their observations, and that means certainly as far back as the ancient Greeks. If gravity were clearly and directly observable there would be nothing for which to thank Galileo and Newton. We could have seen gravity for ourselves.

But Newton and Galileo did discover things about gravity that had not been known, because these things could not be directly observed. Newton in particular, with the description of an invisible force that acts instantaneously and at great distance, removed gravity from direct sensation. This is what the ancients had missed and gotten wrong, because they could not see it. But Newton got it wrong, too, at least according to the theory of gravity we now believe, the general theory of relativity. Gravity is not an instantaneous force. In fact, it's not a force at all. Gravity is the curvature of spacetime, and this is surely not something we can observe. The fundamental mechanism of gravity, the essence of gravity, is hidden.

So, again, what can be observed regarding gravity? There are frequent overly optimistic declarations from scientists and science media to beware of. For example, recent reports of detecting gravitational radiation, a distinctive feature of the general theory of relativity, and radiation from the very early moments of the big bang, got caught up in much observational enthusiasm. "Gravitational waves have been *directly observed*," said one report (O'Neill, 2014). This is surely a stretch of the concept and abilities of direct observation. Seeing, detecting, and interpreting are different activities, and it pays to distinguish them insofar as we will be interested in the process of science and the reasons to believe that scientific claims are true. We can see light, but that does not mean we can see electromagnetic waves. If we could, there would never have been a dispute as to whether light was a wave or stream-of-particles phenomenon. We can see things fall, but that doesn't mean we can see a gravitational field.

Even in the act of seeing, or more generally experiencing, there is an important distinction to track. We can talk about observing some particular thing in the sense of being able to point it out and distinguish it from the background, but this does not mean it has been observed and accurately identified for what it is. For example, astronomical records show that Galileo had observed the planet Neptune and marked it on a chart of celestial objects. But he recorded it as a star, and not a particularly interesting one at that. It wasn't until more than two centuries later that Neptune was observed and identified for what it is, the eighth planet. Galileo saw Neptune, but Johann Galle was the first to see it as a planet, and it is the latter who is credited with the scientific contribution of finding Neptune. So, when we ask about observation and gravity, we will want to know about meaningful observation that can contribute to knowledge and understanding.

It will be important to distinguish as well between the force of gravity and the motion that results, the **acceleration of gravity**. In the language of physics, **kinematics** is the description of motion, in terms of position, velocity, and acceleration. **Dynamics** is the science of the cause of motion, the forces and masses that promote or inhibit acceleration.

Start the analysis of what is observable about gravity with kinematics, the ways things move under the influence of gravity. This is already one step removed from gravity itself, the force of gravity, but it gives us the most immediately experienced effects of the force. There is no doubting the fact that we can and do observe that things fall. We see objects rise and stop and then fall back down. We see the curved trajectories of projectiles. And we see and carefully record the motions of celestial objects.

What goes up must come down. Strictly speaking, and since this is science we really have to speak strictly, what we observe is that what goes up *does* come down. There is a difference in saying that something *must* happen and saying that it simply *does* happen. The first implies a necessity that is missing from the second. And the necessity is never seen. That is, there is a conceptual difference between does and must – we understand the difference – but there is no possible empirical difference. All we see is the sequence of events.

So, what we observe is that what goes up, does come down. But even that is not exactly what is observed. There is an implied universal generalization in the claim, that *everything* that goes up, comes down. That's not what is observed. For one thing, no one of us, and not even us collectively, has observed everything that goes up. We have only a small sample of such events, and all of those have been in the past. There is no evidence of things going up in the future and whether they will come down. So the observation report, again strictly speaking, has to be revised. What has gone up has come down.

But, of course, there are counter-examples. Not everything we have observed to go up has come down. Helium balloons, clouds of smoke, and recently some spacecraft sent beyond Earth's orbit have gone up without coming down. And some things that go up come down but in an obviously controlled way. They don't fall down. Airplanes and hot-air balloons go up, and yes they come down, but in a way quite unlike a tossed ball.

The summary of the observation of this most basic effect of gravity, in other words, is complicated. The direct observation, honest and without inference or amplification, is that what has gone up has usually come down. This is a long way from what we would call a law of nature. It lacks both the universality and the necessity, as in, it's a *law*! But it's what we can observe.

We can observe these phenomena in more detail, and thereby add the precision that facilitates a mathematical description of motion. Aristotle claimed that things fall, and more precisely, heavy things fall faster than light. Galileo famously disagreed, insisting that heavy things and light things fall at the same rate. This seems exactly the sort of disagreement that can be settled by careful measurement, and Galileo is often credited with adding exactly that, careful measurement, as a necessary component of scientific method. But try it. Drop something heavy like a brick and something light like a dry sponge and usually the brick hits the ground before the sponge. So, even though the equal rate of falling is a fundamental principle of our understanding of gravity, the Principle of Equivalence, the most basic observations of freefall do not exactly support the claim. Scientific evidence is not in naïve observation but in considered observation, that is, experience and reason together. And in our case, reason suggests there are more things influencing the falling object than gravity alone. There is air resistance.

In Galileo's time, and today when we do the experiment in a room filled with air, there is just one observable effect but it is caused by a combination of two forces, gravity and air. There is no way to observe the separate influences of the two causes, since they are always at work together. Galileo had to use reason to infer what *would* happen if a heavy object and light objected were dropped together in the absence of air. They *would* fall at the same rate. But just as one can't observe what *must* happen, one can't observe what *would* happen. We only observe what *does* happen. And so this important principle of equal rates of freefall was, when first introduced and for much of its influential run in the science of gravity, not a matter of direct observation. Only now, what with trips to the Moon and very good vacuum chambers, can we remove the impediment of air, and see quite directly that Galileo's reasoning was right.

Separating contributing components within a composite observation requires a ruling on what is relevant to what. Air resistance is not relevant to the fundamental mechanism of gravity, and so its effect must be removed, by reason or by vacuum, from the observed effect. Recognizing relevance presumes some basic understanding of what is going on and what can be ignored. Galileo had to assume that air resistance and gravity were not in some fundamental way connected, and that there were these two distinct forces producing the one observed effect. These sorts of hypotheses about what is and what is not relevant are standard scientific practice.

Galileo went further than the Principle of Equivalence, and proposed a mathematical description of the rate of freefall. This is the so-called time-square law, that the distance an object falls, however heavy or light, is proportional to the square of the time it has fallen. Observing this, or more to the point, measuring this, requires precise timing. It needs a good clock, something Galileo lacked. Things fall really fast. But Galileo slowed them down to the point where he could keep track of where they were at uniform intervals of time by having them roll down a slightly inclined ramp. Instead of a freely falling stone he measured position as a function of time for a slowly rolling ball. Distance down the incline increased as the square of the time elapsed. To use this observation of rolling down an incline to conclude that an object in freefall obeys the time-square law requires some robust reasoning. In other words, Galileo didn't observe the time-square law for freefall. He observed the law for rolling down a ramp, and then reasoned that the two factors that differed between the cases, moving on a slope versus straight down, and rolling rather than flying with no rotation through the air, were irrelevant.

There is something to be learned about scientific observation in general. Genuinely *scientific* observation is not just a matter of objectively taking in the information given by nature. There is much to consider, and hence an essential role for reason in the observing process. Relevant factors must be controlled. Only relevant properties are to be noted, while the irrelevant are to be ignored. When the apple dropped, Newton was not distracted by its being a Macintosh or Delicious, or even fruit or mineral. He attended only to the relevant property of it being an object with some mass. But the key factor of what is and what is not relevant to the phenomenon in question is itself not immediately observable. This is a bit of theory. And this is part of the give-and-take between theory and observation in science.

Turn now to the possibility of observing the force of gravity itself, the dynamics. In this case, we can't expect to see the force, but we might hope to feel it. And it seems we do, since we have already reported being able to distinguish between a heavy thing like a brick and a light thing like a dry sponge. We also experience the pressure on our feet when we stand up or on our backside when we sit down.

Strictly speaking, what we experience in these cases is the force opposing gravity rather than the force of gravity itself. Again, there are multiple forces at work when we hold the brick; there is the force of gravity and the force we supply to hold the brick in place against the pull of gravity. In circumstances in which gravity is the only force on the object, that is, when it is in freefall, we feel nothing. This would be the case of falling with the brick, jumping off a cliff with brick in hand, for example. The brick feels weightless, as do we. We cannot observe the force of gravity, since when it alone is applied to the brick or ourselves, we feel nothing.

Using the laws of physics, we know that if an object accelerates it must have a force applied. So observing the brick fall, that is, accelerate downward, we can say there must be a force on it. This is gravity. Right, but this is an inference, using reason based on the law that acceleration is caused by a force. This is not actually observing the force. Furthermore, to figure out the force on an object based on the measurement of its acceleration, you have to know its mass. Force equals mass times acceleration. Knowing the mass of objects in hand, that is, objects that you can put on a balance and compare one to another, is directly observable. If it takes three dry sponges to balance the one brick, the mass of the brick is three times that of a sponge. But there is no direct way to measure the mass of a celestial object, something like the Moon or the Sun or a planet or a star. These won't go on a balance. Astronomers can tell us the masses of these things, but these data are very indirect, most often relying on a theory of gravity to figure out the mass from observations of how the object moves. There is a challenge, then, in establishing a law of *universal* gravitation. You have to know the mass of a distant object to test the theory of gravity. And you have to use a theory of gravity to know the mass of the distant object. The reason to believe the theory is true cannot be that it is simply based on observational evidence. It's going to be more complicated than that.

And that's the moral of the story in asking what is observable regarding gravity. The science of gravity is not just a matter of generalization of direct observation. We will need to watch the give-and-take between observation and theory to understand the method and the reasons to believe that the results are true, or at least likely to be true.

Despite this complication, scientists are generally quite confident in their current theories about gravity. Confidence in current theory is in fact a hallmark of science, and always has been. Aristotelians were generally so sure they had it right that they considered almost no change or challenge to their theory for over 2000 years. What goes up must come down because the element earth, what makes heavy things heavy, seeks out its natural place at the center of the universe. When theoretical change did come in the form of the Newtonian theory of a universal force of attraction, Newtonians embraced the new description with similar commitment and loyalty. Even when the theory was consistently wrong in predicting the orbital position of the planet Uranus, the fault was not in our theory but in our planets. Sir George Airy, the Astronomer Royal from 1835 to 1881 and as characteristic a spokesman for contemporary science as we could find, declared that everyone was "fully impressed with the universality of [Newton's] law of gravitation." (Quoted in Moore, 1996, p. 94.) There had to be some unseen object pulling on Uranus, and indeed there was, the planet Neptune. But eventually the theory was discarded and today we describe gravity not as a universal force but as curvature of spacetime. And again there is a general attitude that we got it right this time. And again the theory is not matching all the observed data, and again it's not the theory that is challenged. Instead, a new form of matter, dark matter, is hypothesized to settle the mismatch between theory and observation.

The history and current events of the science of gravity show a commitment to the theory on the books, even in the face of challenging evidence. It is surely not the case that a single failed prediction forces scientists to abandon or even revise a theory, contrary to descriptions of scientific method in many textbooks. That is not only a simplistic account of how science works, it would be counterproductive and inefficient. No theory is empirically perfect. To give up on an idea when challenged by just one observation would cause such theoretical instability that no idea would be fairly considered or fairly tested.

The three main theories of gravity that make up the history of the science of gravity, Aristotelian, Newtonian, and the general theory of relativity, have all been carefully considered and accountable to the evidence in their time. In light of what we know now, that is, according to the general theory of relativity, the Aristotelian and Newtonian theories are false. This record of failure, of settling for false descriptions of gravity, should be caution against over-confidence in the current theory. Aristotelian theory seemed reasonable and matched what they took to be the important evidence, but it turned out to be false. Newtonian theory seemed reasonable and matched what they took to be the important evidence, but it turned out to be false. The general theory of relativity now seems reasonable and matches what is considered to be the important evidence, and it is judged to be true.

Aristotelians had no history of prior theories to learn from, no record of failure, so their own confidence was, in the historical context, reasonable. Newtonians had just one historical data point, Aristotle's false theory, so there was no trend, no pattern of failure. But now, the historical record is consistent, a correlation between high confidence and false theory.

Confidence in current scientific theory, in spite of the historical evidence that theories change significantly, is normal. In fact, there are systematic components of the scientific process that make it almost a necessity. One of those is the role of textbooks in the education of a scientist. The other is the authority of peer review as a standard of credibility and acceptance of new ideas. Both of these enforce the network of current ideas and practices, with the implicit attitude that those ideas are right. This network of commitments makes up what Thomas Kuhn called a scientific paradigm.

Consider the role of textbooks in scientific education. Non-scientific disciplines rely on textbooks for the first few years of an undergraduate's training, but soon add in primary sources and journal articles. Graduate students in history or philosophy never use textbooks. As students they participate in the give-and-take of conflicting arguments and interpretations; that's an essential part of their education. But the education of a physicist is from textbooks, start to finish. Graduate classes in quantum mechanics or relativity are based on textbooks on quantum mechanics or relativity. This is noteworthy since textbooks rarely encourage doubt or dispute. Their goal is clarity in presenting the fundamentals, and the tone is confident. This is what we know and what you as a scientist will build on.

And now think about the process of peer review, a cornerstone of acceptability of new ideas and a point of pride among scientists. What makes someone a peer, and hence an appropriate judge of a grant proposal or a paper submitted for publication, is thorough training in the fundamentals of the science. You know the basics since you have been through the textbooks. The review process is, at least in part, a check for consistency with established ideas, again an enforcement of the current theories and again with the implicit attitude that the current theories are true. There is room for skepticism about some details of application of what is currently on the books, but the books themselves are not challenged. The current theory is used with confidence and conviction.

The pervasive authority of current theory is both a strength and a weakness in the scientific process. It is a strength in that it provides a stable foundation on which to build. New ideas are based on old ideas, and the old ideas in textbooks give expert scientists a head start that facilitates progress. But the foundation can be so stable that it makes big changes, paradigm shifts, almost impossible.

Somewhere in this tension between commitment to established ideas and openness to change is scientific method, the dialogue between theory and observation. Laws and theories make it into the textbooks for a reason, through some challenging screening-process based on conceptual clarity and evidential support. The screening process is not perfect, or science would get it right, absolutely right, every time. History shows that it doesn't. But that doesn't make the screening process worthless. Our goal, the part that involves understanding scientific method, is to look carefully at how evidence supports theory and theory influences evidence, to see how the resulting coherence can be taken as reason to believe the theory is true. We won't do this in the abstract, but by seeing what happens in the science of gravity.

2

Forces and Fields

The idea of scientific literacy, a commonly used concept to praise, or lament, or measure the nation's familiarity with science, suggests that understanding science amounts to understanding a language. What counts is your vocabulary. There is surely more to it than that, and a clear sense of how the terms fit together and relate to things in nature will be required for a genuine understanding of science, but learning the terms is the place to start. Once we get the individual ideas in focus, we can work on the connections.

If we are to talk seriously about gravity, that is, talk scientifically about gravity, we'll need to talk with precision and consistency. This is the first necessary condition of scientific method. What goes up, must come down. But we won't make much scientific progress with vague words like up and down, fast and slow, heavy and light. We'll need specified reference frames and shared systems of units in order to sharpen the evidence, communicate the results, and test the hypotheses. We'll also need to work with some new concepts, forces and fields, to describe aspects of the phenomena that are not directly observable. Clarity on these concepts, forces and fields in particular, is the work for this chapter.

First a caution. There is a difference between precision and accuracy. You can describe something with great precision but be wrong. If you promise to show up at the restaurant at 7:19, but don't make it until 7:30, you have been very precise, three-significant-figures precise, but inaccurate. Back off on the precision and promise to be there some time between seven and eight, and your 7:30 arrival makes the prediction accurate. The methodological point is twofold. One should not be impressed by precision alone. It's accuracy in both description and prediction that counts. But precision is valuable in that it sharpens the image of nature and facilitates the testing of a theory. It is easier to see if a theory is wrong when the details of the theory, and its empirical predictions, are precise.

One more caution. Having a term for something, and even being able to apply the term with precision, does not mean that the thing named is real and actually exists in nature. The average American family has 2.1 children. This is precise and currently accurate, but it describes something that doesn't exist. No one has a 0.1 child in the house. There are other examples of scientific terms that are clear and useful but refer to fictitious things. IQ, the measure of innate intelligence, was once a precisely measured and theoretically important property of individuals, but there is now good reason to say it measures nothing real. And one more example, botanists talk about something called plant strategies, referring to aspects of a plant's physiology or behavior that give it a survival advantage in its local environment. But of course plants don't strategize. They don't plan for their future, and to describe what does in fact happen as adaptation is not to suggest things are done on purpose or by design, the plant's or God's. But the concept of a plant strategy is a very useful way of describing and anticipating details of plants in nature. In this case the accuracy of the precise description is beside the point; it's merely the usefulness of the concept that counts toward its contribution to science. This is something to keep an eye on when we introduce and deploy some hard-working concepts in the science of gravity. We'll need to explicitly note the differences between precision, accuracy, and pragmatic value in things like fields, epicycles, and geodesics.

The science of gravity starts with describing particular cases of motion, and it advances by explaining the cause of the motion. These two activities match an important distinction in physics between kinematics and dynamics, both aspects of the more general branch of physics that is mechanics. Kinematics is the description of motion; dynamics is about the cause. What goes up generally comes down; this is a kinematic statement. Asking what makes it come down gets you into dynamics. Kinematics will be in terms of position of the object, its speed, direction of motion, and acceleration. All of these properties are generally observable. Dynamics, at least since Newton, will be in terms of the forces on the object and its mass, the innate property that resists the force. Dynamic properties are generally not observable, at least not directly. Connecting the kinematics to the dynamics is the challenge in the science of gravity, and it is exactly where scientific method is critical. The link between the observable evidence and the unobservable theory is what the method hopes to secure. Our goal is to clarify how this is done.

Start with the kinematics and clarify the key concepts and terms. The object itself, the thing that goes up and then down, and even the planets in orbit, are generally idealized to be described as a single point. The position can be reduced to the position of just this one point. A precise and intersubjective determination of the position, something more focused than simply over there or on the right, requires a **reference frame**. It's our choice where to set this up, where to fix the origin, but once that is done the position of the point-object is given by its coordinates along the x, y, and z axes.

Position is determined only in reference to a coordinate system as described above. Note the two meanings of the word "determined." One sense of determining is our knowing the value of the property. We can't determine the position of an object without a reference frame, a coordinate system as the fixed context. The other sense of determine is not about us or our ability to know a property; it's about the property itself having a fixed value. It's about the physical criteria to give the property a value. The difference between these two senses of determination will be worth keeping track of in both the science of gravity and the description of scientific method, so it's worth making very clear. Here is a voting analogy. At the moment when the last vote is in, the winner of the election has been determined in the sense that there *is* a winner even if we don't yet *know* who it is. This is the physical-criteria sense of determine. But it's not until the last vote is counted that we have determined who the winner is, that is, we have measured the value of this property. This is the knowing-the-value sense of determine.

Back to kinematics. A reference frame is necessary to determine, in both senses, the position of an object. This is because position is relative to reference frame, and it can only be measured with respect to some other object that defines, or at least is itself already located in, the reference frame. And, of course, precise and sharable information on position will have to be in terms of standardized units such as meters, feet, or light years. This is to express the distance along each axis from the reference-frame origin.

If the object moves, and it will, or there's not much to study in the science of gravity, the motion will be in terms of the change in position per time, usually in units of seconds. That is, the velocity v is the distance d travelled per time t: $v = d/t$.

Velocity is a **vector**. The change in position has both a magnitude, a distance, and a direction. Moving 10 meters to the left is a different phenomenon than moving 10 meters to the right, and the description of velocity will have to keep track of the difference. Going 10 m/s up is different from going 10 m/s down. In fact, the property of position is a vector, too. It has the information of both distance from the origin of the coordinate system and direction from the origin. In an x–y coordinate system as in Figure 2.1, the position of point A is determined by its coordinates (1,3). The two numbers are required to specify the position in the two-dimensional coordinate system, supplying the two pieces of information, distance and direction from origin. B is at (4,–1). The displacement between A and B also a vector, since it has both a magnitude and an orientation. We can figure its magnitude, the distance d, by a simple application of the Pythagorean theorem: $d^2 = \Delta x^2 + \Delta y^2$, where $\Delta x = \left(x_B - x_A\right)$ and $\Delta y = \left(y_B - y_A\right)$. In Figure 2.1, $d^2 = (4 - 1)^2 + (-1 - 3)^2 = 3^2 + 4^2 = 25$. So, $d = 5$.

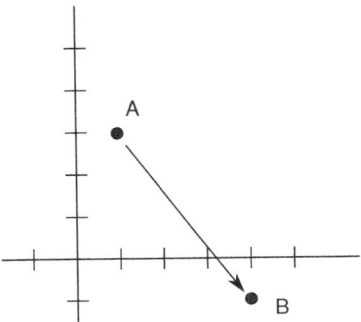

Figure 2.1. The vector displacement between two points A and B. The distance is calculated using the Pythagorean theorem.

The velocity in moving from A to B will depend on the amount of time it takes to get from one point to the other, and on the units of both distance and time. If the units in the coordinate system are in meters, and the trip took, say, 0.5 seconds, then the velocity is 10 m/s, in the direction of *d*, as shown. The magnitude of the velocity, the 10 m/s in this case, is referred to as the speed.

Properties that are vectors will be identified as such by being printed in bold. Thus, **v** = 10 m/s, in the direction indicated by the arrow in Figure 2.1.

If the velocity changes in any way, the object speeds up or slows down or turns, this is acceleration. Any change in a vector quantity is itself a vector, as change in position (a vector) is velocity (a vector). So, acceleration is a vector. Acceleration **a** is the change in velocity per time. If the units of position are meters and the units of time are seconds, then the units of velocity are m/s and the units of acceleration are (m/s)/s, or, m/s^2.

In equations, vector quantities are written in bold. In diagrams, vector quantities are presented as arrows. The length of the arrow corresponds to the magnitude of the vector, and the direction is, naturally the direction. So, in Figure 2.2, if $\mathbf{v_1}$ represents a velocity of 10 m/s to the right, $\mathbf{v_2}$ is the velocity 5 m/s at 45° in the *x–y* plane.

Not all important properties are vectors. Temperature, for example, has no direction. You wouldn't say that water boils at 100 °C to the right, or up. It's just 100 °C. Similarly with the mass of an object. There is a magnitude, 60 kilograms, say, but no direction. So, these properties are not vectors; they are **scalars**. In these cases, precision requires standardized units and a reference, but just one number will determine the value of the property. Temperature can be measured and reported in Celsius, a system that has the freezing point of pure water as its origin, its reference for 0°, and a fixed unit to measure degrees. Just as a coordinate system to determine position, velocity, and acceleration requires a physical object as a

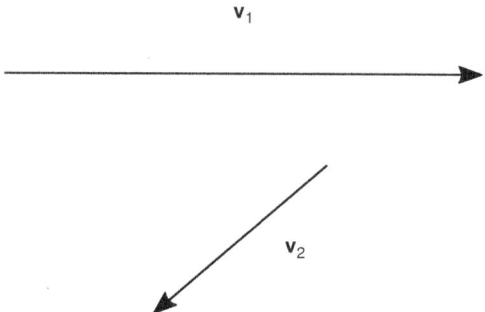

Figure 2.2. Two velocity vectors. $\mathbf{v_1}$ represents a velocity of 10 m/s to the right. $\mathbf{v_2}$ represents 5 m/s at an angle of 45° down and to the left. $\mathbf{v_1}$ is twice as long as $\mathbf{v_2}$ to indicate that it has twice the magnitude, that is, twice the speed.

reference point, this thermal system requires a physical phenomenon, water freezing, as a reference. Scalar quantities like mass are written into equations in a normal, not bold, print, as in the mass m. There is no arrow associated with a scalar.

These are the tools we need for a precise description of motion, the science of kinematics. If we had these numbers, units, and arrows on the blackboard, anyone looking in would be able to tell it's a science class. But we're not really doing science yet, since we haven't really described anything in nature yet. We have the words but we haven't constructed any sentences. The next step, when we'll start doing science, is to make connections to and among the terms, connecting them individually to things in nature and connecting them to each other. The distance formula, the Pythagorean theorem applied to Figure 2.1, was a start. It said that $d^2 = \Delta x^2 + \Delta y^2$. This relation between distance and coordinate positions is true for any two points in any Cartesian coordinate system. This sort of generalization is exactly what to expect from science. It's not just a precise description of a particular situation; it's a universal formula that applies to all such situations. It's a law.

Now that we have the vocabulary of kinematics, the concepts of vectors and scalars, and the will to generalize, we can turn to the concept of a **field**. The most general characterization is that a field is a physical parameter that depends on position. It is a smoothly varying, continuous array of parameter values. This is perhaps too general and abstract to help, so consider some examples of fields.

The distribution of air temperatures at all points across the country is a field, a temperature field. A weather map in the newspaper shows a few selected values of the field at points of interest, but the field itself is continuous. There is a temperature at every single point in the country, measured or not, reported or not. Furthermore, there are no abrupt, discontinuous jumps in temperature from one point to its neighbor. If you start in Miami at 30 °C and travel to New York where it's 8 °C, you will pass through all of the intermediate temperatures between 30 °C and 8 °C. This is not a force field, since temperature is not a force. And describing the parameter as air temperature does not mean that it's the air that's the field. The air in this case is the medium, the thing that bears the property. It's the temperature itself that is the field. And since temperature is a scalar property, this is a scalar field.

The weather report could also include a map of the wind field. This would show the speed and direction of the wind at various points around the country. The field itself covers every single location, reported or not, though some values of the parameter would be zero, places where the air is dead calm. This is a vector field, since wind has both magnitude (speed) and direction. At each point in the wind field there is a vector pointing in the direction the wind blows. The length of the vector is proportional to the wind speed. Again, the moving air is not the field; it's the medium. And again this is not a force field, since speed is a purely kinematic property, but we're getting close to the idea of a force field. Wind speed is related to the force you would feel if you were there, as in gale-force wind.

To connect the kinematics of wind speed to the dynamics of force, consider another scalar field on the weather map, the field of atmospheric pressure. Pressure is the force per area, as in pounds per square inch. Since force is involved, this is a dynamic property. But despite the involvement of force, pressure is a scalar. It has no direction. It's the force in all directions at some point, calibrated per area. So, the pressure field is a scalar dynamic field. It's a good example to show why the word dynamic is used to describe this kind of property. Pressure will make things happen.

One point in the country might report an atmospheric pressure of 1018 mb. This is 1018 millibars. Millibars are the unit of pressure preferred by meteorologists, and 1 bar, that is, 1 b, corresponds to a typical atmospheric pressure. 1000 mb = 1 b, so our example of 1018 mb is not unrealistic. Different points in the country may have different values of pressure, just as they had different values of temperature and wind velocity. The array of pressure values is the field. And, just as with temperature, the change in pressure values is continuous as you move from one point to another.

Now add a feature to the pressure field by drawing lines that connect points that have the same value of atmospheric pressure. This is clearly an artificial addition to the map, and there are no real lines across the country, but it will be a very useful addition. These lines are called isobars, since they are lines along which there is no change in pressure. They are like contour lines on a topographic map, lines that run through points at the same elevation. You could also construct the isotherms on the map, the lines that connect points that have the same temperature, but for our purposes it's the isobars that will be more informative. This is because they are dynamic. They will reveal connections between forces and motion.

An example of a portion of a weather map with isobars is shown in Figure 2.3.

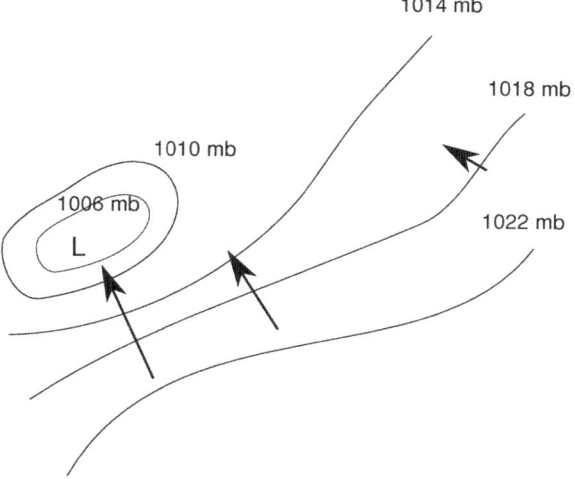

Figure 2.3. A weather map with isobars and wind-velocity vectors. The wind blows from high to low pressure, faster where the pressure gradient is stronger, that is, where the isobars are closer together.

Note that the isobars in the figure never cross. This is not an accident, nor is it a feature peculiar to this example. Isobars can never cross, and this is an important feature of the field. If the isobar for 1018 mb crossed the one for 1022 mb, the point of intersection would have two values for the one property, and that's impossible. All points on the map have exactly one value of air pressure, so the lines can neither cross (indicating two values) nor break (indicating no value). The lines can curve around and form closed loops, as the 1010 mb and 1006 mb lines do in the figure. This is analogous to closed contour lines representing a hilltop or a valley. In Figure 2.3, the closed region within the 1006 mb isobar is a region of low pressure, or simply a low, in the language of weathermen. It's marked by the capital L.

Air flows from an area of high pressure to low, just as a ball rolls downhill, from a position of high elevation to low. And the ball rolls faster if the grade is steeper. So too does the wind blow faster if the gradient, the rate of change, from high to low pressure is steep. With this information, we can look at the map with isobars and know things about the wind in that part of the country. Wind will not blow parallel to an isobar, any more than a ball will spontaneously roll on level ground, that is, along a contour. Wind will always blow perpendicular to an isobar, and always from high to low. And where the isobars are close together, where the gradient of pressure change is steep, the wind will blow faster than where the lines are far apart.

The arrows in Figure 2.3 show a few values of the wind velocity. They are not random. Rather, they correspond to the details of the isobars in that they always point toward the low pressure, and they are longer, indicating faster wind speed, where the isobars are closer together. There is this physical connection between the pressure field and the wind-velocity field. This is science, not only describing the physical situation in precise and consistent terms, but finding and using the connections between one aspect of nature and another. In this case it's a connection between kinematics and dynamics, between the wind velocity and the pressure. The scalar dynamic field of air pressure has information on the vector kinematic field of wind velocity.

The field concept will be very useful in a variety of situations, particularly where we're looking for connections between kinematics and dynamics, gravity, for example. So, the next step is to extend the idea to other aspects of physics. And this is another characteristic technique of science, using analogies, reasoning that one thing x is like another thing y, so y-laws probably apply to x. We've done a bit of this already, for the purpose of illustration. Isobars are like contour lines, and things move downhill perpendicular to the contour lines, so wind blows down-pressure, perpendicular to the isobars. Using more analogies, we'll extend the field concept to get the most out of its descriptive potential.

Most of us have an intuitive understanding of an electric or magnetic field, and these are closely analogous to our eventual target, the gravitational field. When there's electric charge in play, we know that opposites attract and like charges repel. More precisely, as it is the goal of this chapter, there is a force between any two electrically charged objects. The direction of the force depends on the sign, positive or negative, of the two charges. The magnitude of the force depends on the magnitudes of the two charges, q_1 and q_2, and on the distance r between them. The force decreases as $1/r^2$, as the distance r increases. The force is a vector, since it has both magnitude and direction, and it is, of course, a matter of dynamics. But it's not a field, yet. There are just two points in space, the locations of the two electrically charged objects, with values of the force. A field requires values of a property at all points in space, as the temperature field had values of air temperature for every point in the country, not just New York and Miami. But the $1/r^2$ dependence shows a continuous function of position, at least position relative to one of the charges, and this is what we need to determine a field.

Using one of the charges q_1 as the focus of our attention and the origin of a coordinate system, ask what the electric force *would* be on the other charge q_2, if q_2 were at some different point in space. Charge q_2 doesn't have to actually be at that other point for the force to have a determined value, and this is true for every point in space. There is nothing really there, at those points in space; there is only the one object with charge q_1, at the origin. There doesn't have to be air or any other kind of medium. The laws of electrostatics determine, for every empty point in space, the force q_1 *would* produce if another charge, q_2, *were* at that point. This

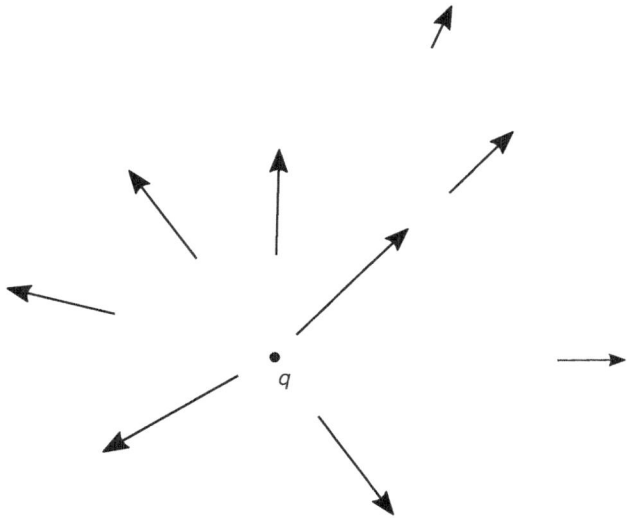

Figure 2.4. Electric-field vectors near an electric charge q. The field always points directly away from q, and the strength decreases further away from q.

is the field, since there is now a determined value of the force at every point in space, a value that changes continuously from one point to the next. To make the value consistent from one theoretical description to another, we need to decide on a fixed value for q_2. This is a matter of convention, like choosing which units to use in measuring temperature or distance. The so-called test charge, positive by convention, is then the charge that, if it were at some point x, would feel the force from q_1, given the distance between q_1 and x.

Thus, any single electric charge will be the origin of an electric field. The field can be represented by drawing in some of the vectors at some of the points in space. The field of a single positive charge q will have the vectors all pointing away, along radial lines from q, since like charges repel, and the test charge is stipulated to be positive. The field vectors will get shorter for points further away from q, since the magnitude of the force decreases with distance from q. This is shown in Figure 2.4.

Since the vectors in Figure 2.4 represent the electric force, this is a dynamic force field. That looks and sounds impressive, almost intimidating, but it's worth repeating that there is nothing there in the space surrounding q, no other charges, no medium, and, unless another charge is brought into the picture, no actual force. The field can and will make things happen, it's dynamic in this sense, but only when that other charge shows up.

Despite this immaterial nature of the electric field, it has some genuine dynamic properties on its own. If the source q is changed in some way, by moving it or changing the magnitude of its electric charge, the nearby field vectors change. They

change direction and change magnitude accordingly. This process of alteration in
the vectors moves through the field at a finite speed, in fact, the speed of light.
Thus, the field is at different stages of change at different points in space. The
effects of changing the source move through the field, and in time the test charge
will be affected by this different force and react accordingly. Moving the source q
will eventually cause the test charge, or a real charge if there is one there, to move.
In this way, the energy it takes to move q will be transferred to the distant charge.
In the time it takes between moving q and the field reorienting at the position of
the distant charge, the energy in transit from q to that distant charge is in between
the two. The energy is in the field itself. It can't be at the distant charge, because
nothing has happened there yet. And it can't be at q, since its motion has stopped.
The energy must be in the field, moving from q outward. But the test charge could
be at any point in space, so there must be energy in the field moving out from q in
all directions.

The change in the field and the energy it carries are called electromagnetic radiation. The radiation moves through space, empty space, as a wave pattern in the field
lines. It moves at the speed of light, fast but finite. There is energy, the potential to
make things move, but no medium.

It's hard to know whether the concept of an electric field refers to something real
in nature, or is merely a useful construct like plant strategies and 2.1 children, for
which there is no corresponding object or property. We started out with what seems
like a very unreal construction. What *would* be the force if a test charge *were* at point
x in space. Not only is there no test charge, there is (consequently) no force. There
is nothing, other than the source charge q. Drawing in the vectors and field lines is
then prescribed by the mathematical details of the theory of electromagnetism, but
it is an act of drawing in. There are no lines to discover in nature. And describing
the energy, the radiation, the waves, as moving shapes in the field lines is giving
dynamic detail to features we have put into the picture ourselves. The field lines
are, as a matter of principle, unobservable, since they are properties of nothing. All
of this suggests that the field lines, and the electric field generally, are a helpful
metaphor but not the true literal description of what's going on in nature. You will
never observe isobars in nature either, not as actual lines that you could photograph,
but at least there is *something* there, the air, that is the object bearing the property,
the air pressure. Isobars have a better claim to reality than electric field lines.

But it's hard to dismiss the electric field as unreal. Energy gets from one charge
to another, and the energy is located somewhere between the two charges. If energy
is a property, the ability to make things move, it must be a property of something.
There is nothing material between the two charges, so the something that bears the
property, the energy, must be immaterial. That's the field. The field has a causal
capability to affect any charge anywhere. It could kill you, if the frequency of the

wave was high enough and the intensity strong enough. These would be gamma rays, destructive in a real way with real energy. It's not a metaphorical death, so it couldn't be a metaphorical field.

Perhaps asking the question about the reality of the field is stepping outside the boundaries of science, across the border into philosophy. The more pressing *scientific* question is probably just whether the field concept is useful. Is it worth the abstract trouble? The answer to this one is easy. Yes. Understanding modern theories of gravity will require the application of the concept of fields. Newton's theory of gravity was seen to be problematic, even to some extent by Newton himself, in that it involved a dynamic interaction between two massive objects, the Sun and a planet, say, or the Earth and an apple, without any material contact between them. It also allowed this interaction to be transmitted instantaneously, across whatever large distance between the objects. This so-called **action at a distance** was the conceptual downfall of Newtonian gravity, and it was replaced by a field. A source of gravity creates the field and interacts with the most immediate points of the field, while a distant object is influenced by the field at its location. All interactions are between things and fields. The field obviates the need for action at a distance. The field is an important component of the gravitational interaction. It's useful to the point of being indispensible.

We switched from asking whether fields were real to whether they were useful, but maybe there is a connection between these two virtues. How could a concept be of such great theoretical and interpretive value if it did not correspond to something in nature? Germ theory is such a help in keeping us healthy because there really are germs and they really are washed away with soap and water. How could the theory succeed if it was built on fiction? Keep this question in mind as we see just how successfully fields are used to describe the phenomena of gravity. The question is whether this success is an indication the theory is true.

Modern theories of gravity employ fields very much like the electric field. There is a source but no medium. In the general theory of relativity there is even radiation; there are gravitational waves in the field. The source in the case of the electric field is the electric charge. There is an object, an electron or rubbed balloon, but what really matters is the property of the object, its electric charge. Only some things have electric charge, some positive and others negative, and only these create an electric field or respond to an electric field. Charge is the property that couples these kinds of things together by causing a mutual force between them. Charge couples to charge, not to mass or color or any other property of an object. The coupling force **F** is proportional to the two charges and inversely proportional to the square of the distance between them.

$$\mathbf{F} \propto q_1 q_2 / \mathbf{r}^2 \tag{2.1}$$

The symbol $\propto$ is read as "is proportional to." There is a constant of proportionality, a numerical fact of nature, that would turn the $\propto$ into an $=$, but for now all we need to keep track of is the proportionality. This shows how the important properties contribute to the force. This expression is pure dynamics, in that it describes only the source of action, the force, without any indication of the effect, the resulting motion.

The resulting motion is the kinematics, and it is Newton's second law of motion that links the dynamics of force to the kinematics of motion. For any kind of force $\mathbf{F}$, regardless of the source, the result is a proportional acceleration $\mathbf{a}$: $\mathbf{F} \propto \mathbf{a}$. In this case we know the constant of proportionality; it's the inertial mass m of the object.

$$\mathbf{F} = m\mathbf{a} \tag{2.2}$$

In the case of two electric charges, the force is caused by their charges but their reaction to the force, their acceleration, is proportional to each mass. If the force between two objects is caused by a coiled spring between them, the strength of the force is determined by the characteristics of the spring, but again their acceleration is (inversely) proportional to each mass. The force of gravity will be different from these other cases in an important and interesting way. With gravity, the force is caused by the masses of the two objects, and, as with all other forces, the acceleration is inversely proportional to the mass. In other words, the mass of an object plays a role in both cause and effect; it's on both sides of the equation. This is unique to gravity. It's part of what makes gravity both enigmatic and ubiquitous.

Newton's second law connects dynamics to kinematics, force to acceleration, and whatever the nature of the force, the connecting property is the mass of the object. Knowing this, we are able to use observations of how things move, kinematics, to draw conclusions about why they move, dynamics. We can carefully note the motions of things like planets, the Moon, and projectiles, and use these data as evidence for a theory of gravity. This is how Newton claimed to arrive at his theory of universal gravitation. So it's to that theory we turn first, presenting it in detail in the next chapter, and then seeing where it came from in the four chapters that follow.

Saying that the observations of kinematics are the foundations for the theories of dynamics risks a misleading simplification of scientific method. It's not strictly observation first and then theory. Nor is the logic a one-way stream of information from the empirical to the theoretical. No observational data are innocent of some theoretical interpretation and selection. Theory and observation are always in reciprocal influence and conversation. And, of course, there are essential tasks for evidence to come after a theory is proposed. The theory must be tested. Testing Newton's theory, using evidence that was gathered after the theory was on the books, was supportive at first but ultimately revealed flaws significant enough to demand rejection of the theory. We will see Newtonian gravity rise and fall. It will be replaced by the theory of relativity, and we will watch relativity rise. It, too, may eventually fall. What goes up . . . Even the basic language we use today to describe the phenomena and cause of gravity may change. But for now, we'll speak the language of forces and fields.

3

Basic Newtonian Theory

Every physics textbook used for the introductory survey has a section on gravity. It is invariably about Newton's law of universal gravitation, and not much else. The Newtonian theory is what you're expected to know, if you claim to know anything about gravity. It is also the common standard of comparison for other theories of gravity, as in, the Aristotelian theory is unlike the Newtonian in that . . . , or, the general theory of relativity describes gravity differently than the Newtonian theory by saying that . . . For all these reasons, it's a good idea to start the science of gravity with Newton. For most of us, this is starting with the familiar, and making sure the things we have seen before are clear and correct. Familiarity can accommodate complacency and sloppiness, that is, imprecision, and that won't do for a science of gravity.

The presentation in this chapter will be in the straightforward style of a textbook, with little philosophical or methodological reflection. There will be no historical context, nothing about the process that led to or followed Newton's theory of universal gravitation, and no challenge about its being true or false or about things that are real or metaphorical. It will just be the nuts and bolts of the theory. This is pretty much what textbooks do. The goal is clarity in providing the stable, reliable background knowledge that a scientist needs to participate in the current paradigm. It's the important beginning of the expertise and authority it takes to be a peer, qualified to do peer review.

The central concept of Newtonian gravity is the description of the force. This is pure dynamics. The force of gravity between two objects is proportional to their masses m_1 and m_2, and inversely proportional to the square of the distance r between them. That is,

$$\mathbf{F}_{\text{gravity}} \propto m_1 m_2 / \mathbf{r}^2 \tag{3.1}$$

The coupling property in the case of gravity is mass. This is the analog to electric charge; it is the property of an object that makes it both cause and response to the force of gravity. The constant of proportionality, the coupling constant, is called the gravitational constant G. Its value is discovered by empirical measurement rather than by derivation from other properties. It is simply an unexplained feature of nature.

$$G = 6.67 \times 10^{-11}\,\mathrm{N\,m^2/kg^2}$$

The N stands for Newtons, a unit of force. Putting all the pieces together we get the exact formula for the gravitational force between the two objects.

$$\mathbf{F}_{\mathrm{gravity}} = Gm_1m_2/\mathbf{r}^2 \tag{3.2}$$

There are some important details implicit in the formula that should be made explicit.

First of all, the force of gravity is always attractive, unlike the electric force that can be either attractive or repulsive. Mass m has only one sign, always positive. Like masses attract, and all masses are alike in sign, so all masses attract.

Second, the coupling property m, the analog to q in the electric force, is the same property m in Newton's second law of motion, $\mathbf{F} = m\mathbf{a}$. So, with m in the formula for $\mathbf{F}_{\mathrm{gravity}}$, there is an m on both sides of the $\mathbf{F} = m\mathbf{a}$ equation. It's the same m in the cause $\mathbf{F}_{\mathrm{gravity}}$ as in the effect, the acceleration $\mathbf{a}$. It's worth asking whether this is just a coincidence, another basic fact of nature that has no explanation. It is what it is. Or is there a deeper, more fundamental reason for this, and hence an explanation? The double-dipping of mass explains why everything on the Earth, heavy and light things alike, all fall to the ground at the same rate. Well it explains insofar as a coincidence itself can be cited as an explanation. We need to keep an eye out for a reason why what is called the gravitational mass, the m in the $\mathbf{F}_{\mathrm{gravity}}$, is identical to the inertial mass, the m in the $\mathbf{F} = m\mathbf{a}$ formula. But for now, just take it as an empirically well-established fact.

A third noteworthy feature of the formula for the force of gravity is the $1/\mathbf{r}^2$. This is exactly like the electric force, an inverse-square force. The force gets weaker with distance, and quickly, given the *squared* $\mathbf{r}$. It gets weaker but it never goes to zero. This shows gravity to be a ubiquitous glue. It is an attraction between any two objects located at any two points in the universe. The attraction quickly becomes, as the physicists say, negligible, but it is always there. And since the magnitude of the force is a continuous function of position, position of one of the objects with respect to the other, we are on the way to describing it as a field.

It's worth noting a few numerical values of the magnitude of the gravitational force, to put things into quantitative perspective. It's easy to calculate the gravitational force between the Sun and the Earth. Look-up and put in the numbers for the mass of the Sun, the mass of the Earth, and the distance r between the two bodies, and do the math. The result is 3.6×10^{22} N. For comparison, that's about

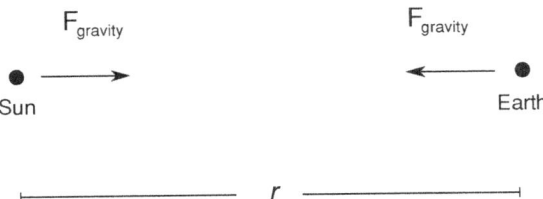

Figure 3.1. The equal and opposite gravitational forces between the Sun and the Earth. The force on each body is directed toward the other. The magnitude of the force is exactly the same on both the Sun and Earth.

8×10^{21} pounds. This is the force of the Sun on the Earth. The Earth weighs 8×10^{21} pounds in the gravitational pull of the Sun. What is the force of the Earth on the Sun? Exactly the same value. The formula for the force of gravity tells us the force of one object on another, without distinguishing which is the one and which is the other. This is exactly in keeping with Newton's third law of motion, that forces come in equal and opposite pairs. The Earth attracts the Sun exactly as forcefully as the Sun attracts the Earth. The Sun is bigger and more massive, but no more forceful than the Earth. The gravitational glue is a mutually shared property.

The formula reports the magnitude of the gravitational force. There is also the direction of the vector to keep track of, but this is easy. It is always attractive and exactly along the line between the two objects. It is a radial force. There is no tangential component of the vector, that is, no component perpendicular to the line connecting the two objects.

This description of the direction of the force vector would be ambiguous for an extended object like the Sun or the Earth or even an apple, since there is not just one point to which the vector could point. Physics deals with this by simplifying the sizable object as if all the mass was located at just one point. The terminology is to regard the Sun and the Earth as point particles. This is not fiction or idealization of mere convenience, since the force of gravity in fact does act as if all the mass is located at just one point. For a sphere of uniform density, the point is, not surprisingly, the geometric center of the object. If the object is oddly shaped or of asymmetric density, there is nonetheless a real point, the center of mass, that is the focus of gravitational force.

With this simplification, the gravitational situation of Sun and Earth can be represented by the simple vector diagram as shown in Figure 3.1. There is an equal and opposite force on each object, directed exactly toward the other.

This two-body situation can be generalized to determine the gravitational field. As in the case of electricity, it will be the field generated by and surrounding just one of the objects, this time one object with mass. And again, the field will be in terms of what gravitational force the object *would* cause at every point in space. The

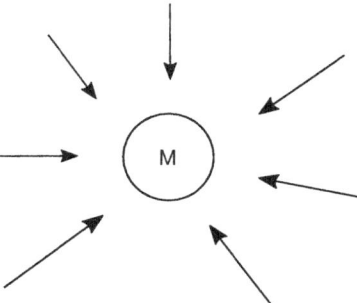

Figure 3.2. Gravitational-field vectors near a massive object M. The field always points directly toward M, and the strength decreases further away from M.

field is determined at every point, even at empty points in space. The units of the field can be standardized by expressing it as force per mass, $\mathbf{F}/m$. This is simply the acceleration of a test mass, determined at every point. In fact, it's the acceleration of *any* mass at that point. This generalization follows from the formula for the gravitational force. To find the gravitational field produced by an object of mass M, now using the capital M to represent not a variable mass but a particular mass, first find the force on a test mass m: $\mathbf{F} = GMm/r^2$. The force per mass $\mathbf{F}/m$ divides out the m, leaving GM/r^2, the acceleration of gravity as a function of distance from the object M. This is the field, with the acceleration vectors all pointing directly to M, getting shorter as they get further from the source. A few of the field vectors are shown in Figure 3.2.

If the source of the gravitational field is the Sun, then M is the mass of the Sun M_{Sun}. Put the Earth in its place a distance r (not bold, since it is just the magnitude of the distance we are working with here) from the Sun, and the Earth will accelerate toward the Sun at a rate of GM_{Sun}/r^2. When you put in the numbers, the result is 6×10^{-3} m/s, a modest number, but acceleration nonetheless.

This raises two questions. If the Earth is accelerating directly toward the Sun, why doesn't it get closer to the Sun? And if the force of the Sun on the Earth has an equal and opposite force of the Earth on the Sun, why isn't the Sun accelerating?

We should answer the second question immediately, because the question itself is mistaken. The Sun *does* accelerate toward the Earth. The force on the Sun is equal to the force on the Earth, but the resulting acceleration is much less because of the Sun's much greater mass. Acceleration is inversely proportional to mass, and the Sun is 3.3×10^5 times more massive than the Earth, so its acceleration is 3 $\times 10^{-6}$ that of the Earth. There is also the complicating factor of the other planets that create gravitational forces on the Sun. Jupiter is particularly influential, because it is so massive. The Sun is pulled in many changing directions, and its resulting motion, its acceleration, is in response to the composite of all these planetary forces.

The answer to the first question, the one about the acceleration of the Earth, opens the discussion of planetary motion, the kinematics of orbit. Keeping something in orbit requires an inward, central force. If you whirl a ball on a string around over your head, you can feel the pull. More to the point, you have to supply the pull to keep the ball from flying off, out of orbit. Any turning, as the ball is constantly turning, requires a force toward the center of the curve. When you drive around a bend, the force comes from friction between the road and the tires. If there's ice on the road this friction is lost and you can't make the turn. This inward force is called a **centripetal force**, regardless of what causes it, whether it's tension in the sting, friction on the road, or gravity. It's always a central force, pointing to the center of the curve, and gravity has exactly that feature. Gravity is the centripetal force that holds the planets in orbit around the Sun. It also holds man-made satellites in orbit around the Earth.

We know the direction of the gravitational force vector is right for the job, always pointing to the center of the orbit, but what about the magnitude of the force? It is possible to mathematically prove that the magnitude of the gravitational force is exactly the centripetal force required for orbit, but it will be better to get the result by a conceptual argument. This way it will make sense.

The faster the orbiting object is going, the more force it takes to make it turn. This is why you slow down to make a turn on ice, as there is less friction, less centripetal force, to pull the car inward and around the bend. This is also why, if you spin the ball on the string too fast, the string breaks. The string just can't handle the tension needed to supply the centripetal force. So expect the centripetal force to be proportional to the speed of the object in orbit. In fact, it's proportional to the square of the speed, indicating that a little more speed requires a lot more force. So, $F \propto v^2$. (Again, these variables are not bold, since they represent only the magnitudes of each property.)

The more massive the orbiting object the more force required to hold it in orbit. If you spin a bowling ball on the string, rather than a tennis ball, it will take more force on your part, and more tension in the string, to hold it in orbit. So, $F \propto m$.

Making a tighter turn requires more force. You're more likely to skid on a sharp turn than a gentle turn. In orbit, the smaller the radius r of the circle, the sharper the turning, so small r requires big F. In other words, $F \propto 1/r$.

Put these pieces together and we have the formula for centripetal force, the force needed to keep a mass m in orbit of radius r, if the speed of the object is v.

$$F_{\text{centripetal}} = mv^2/r \qquad (3.3)$$

This results in an acceleration, since any force causes an acceleration $a = F/m$. This is **centripetal acceleration**, along the line of the force, directed toward the

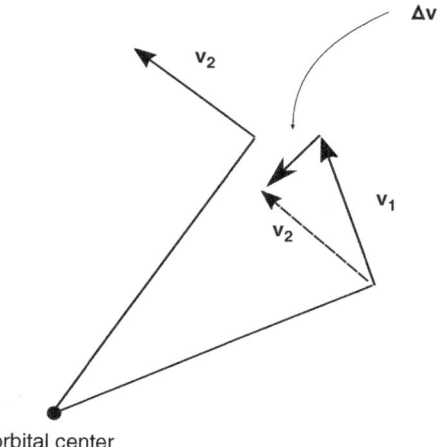

orbital center

Figure 3.3. Two positions in the orbit of an object. The velocity changes from $\mathbf{v}_1$ to $\mathbf{v}_2$. The speed remains constant, so the lengths of the two vectors $\mathbf{v}_1$ and $\mathbf{v}_2$ are the same, but the direction changes. $\mathbf{\Delta v}$ represents the change in the direction of the velocity. This change in velocity is the acceleration, the centripetal acceleration. It points toward the orbital center.

center of the orbit.

$$a_{\text{centripetal}} = v^2/r \tag{3.4}$$

An object in orbit is constantly accelerating, changing velocity in direction but not magnitude. Its speed remains constant, since its speed is always along the tangent of the circle, as in Figure 3.3. The acceleration of the Earth toward the Sun is a matter of changing direction, always being steered toward the Sun to stay in orbit. The Earth doesn't speed up or get closer to the Sun, just as the whirling tennis ball is neither speeding up nor getting closer to your hand. This is the nature of circular motion.

Now put the gravity of a central body M as the source of the centripetal force on an orbiting body m. That is,

$$F_{\text{gravity}} = F_{\text{centripetal}} \tag{3.5}$$
$$GMm/r^2 = mv^2/r \tag{3.6}$$

Note the *m* on both sides of the equation, and reduce the r^2 to get

$$GM/r = v^2 \tag{3.7}$$

This describes the requirements for any orbit of anything, planet, Moon, satellite. It will be so useful throughout the modern science of gravity that we'll give it a name. Call it the **orbit equation**.

For a particular M, for example the Sun or the Earth as the center of an orbit, there are only two variables in the orbit equation, v and r. This shows that there is a fixed relation between the speed and radius of an orbit. Orbit at a particular distance r requires a specific speed v. The distance to the Moon, for example, is 3.8×10^8 m. This is the distance from the center of the Earth to the center of the Moon, and that's the r to put in the orbit equation. There is only one possible orbital speed at this distance. From the orbit equation, that speed is 1×10^3 m/s. Notice there was no need to know the mass of the Moon, since the orbit equation works for anything, independent of its mass. This is one of the wonders of gravity. Put a marble in the same orbit as the Moon and it will have to go exactly as fast as the Moon. The Earth's force on the two objects, marble and Moon, is very different, but the acceleration is the same. The dynamics depend on the mass of the orbiting object, but the kinematics do not. This, we will discover, poses a challenge to figuring out the mass of a distant astronomical object, using observations of its orbit.

The challenge can be met in some circumstances. For example, if M is very much greater than the mass m of the orbiting object, as the mass of the Sun is much larger than the mass of the Earth, we can determine the mass of M from the orbit of m. This is from the orbit equation. Do it for the Sun–Earth orbital system and calculate the mass of the Sun: $GM/r = v^2$. Solve for M. Recall that r is the Earth–Sun distance (center to center), and v is the distance the Earth travels per time, that is, the circumference of its orbital circle divided by one year, the period of orbit. The result is $M_{\mathrm{Sun}} = 2 \times 10^{30}$ kg. In this way you could also find the mass of the Earth, using the orbital details of the Moon.

There are two noteworthy limitations to this technique using gravity and orbits to determine mass. The simple formula only works under the assumption that the central object sits still. That never actually happens, since, by Newton's third law, there is always a force on *both* objects, M and m, so M must be accelerating. We have neglected the acceleration of M by assuming it is so small as to make little difference, M being so much larger than m. The spirit of this chapter is *basic* Newtonian theory, and this shows just how basic the analysis is at this point.

A second limitation is more profound and more to the concerns of scientific method. Any theory of gravity will have to be tested, and that will require knowing the masses of distant objects – it's a theory of *universal* gravitation, after all – to see if the formula for F_{gravity} matches the observations. But we've used the theory of gravity to determine the masses. It would be circular reasoning to use a theory to supply key parameters used in its own testing. We'll have to deal with this when the Newtonian theory, and the general theory of relativity, are put to tests. For now note that it's one example of a more general challenge in testing theories about dynamics. Dynamics are hard to verify; kinematics are easy.

For objects in hand, things on the Earth like apples and stones, we can measure the mass directly, independent of any theory of gravity. So, testing the theory of gravity on the Earth avoids the problematic circularity. As this is a theory of universal gravitation, the formula must be the same on the surface of the Earth as in the empty space of the Solar system: $F_{\text{gravity}} = GMm/r^2$, where M is the mass of the Earth, m is the mass of the apple or stone or what have you, and r is the radius of the Earth. The gravitational field of the Earth is as if all of the Earth's mass is at the one point, the center of the Earth. The distance between the apple and the source of gravity is the distance between the apple and the center of the Earth. This is r.

The force of the Earth's gravity on an object is its **weight**. From the formula, the weight is proportional to mass m. And from Newton's second law, the force will result in acceleration $a = F/m$, that is, $a = GM/r^2$. This is the same acceleration as for an object in orbit, as it should be, since this is universal gravitation, but for an object dropped on the surface of the Earth there is no tangential velocity. In this case the object accelerates, picks up speed, heading for the source, the center of the Earth. Regardless of the mass of the object, this acceleration is 9.8 m/s^2 on the surface of the Earth, the so-called **acceleration of gravity**, abbreviated with the letter g. Drop an apple or a stone and it speeds up at a rate of $g = 9.8$ m/s^2 as it falls. Toss the apple straight up and it slows down at a rate of 9.8 m/s^2 until the speed is zero at the top of its flight, and then it speeds up at 9.8 m/s^2 as it falls. Since this is independent of the mass of the object, heavy things and light things fall at the same rate. As with the orbit equation, the dynamics, in this case the weights, are different but the kinematics, the acceleration and trajectories, are the same. All of this is assuming that gravity is the only force in play. It ignores, for example, the force of air resistance. Since there generally is air in the room when things are dropped, this equal acceleration for all will be difficult to demonstrate.

The kinematics of free-fall are the same for any mass m. They are also the same at any place on the surface of the Earth, assuming the Earth to be a perfect sphere of uniform density. The kinematics are the same wherever you set up your coordinate system. You could even set it up in a train car. Drop something in the train car and it falls, accelerates straight down at 9.8 m/s^2. We didn't specify the mass of the object. We didn't even say whether the train was moving or not. As long as it's not turning, slowing down or speeding up, the law of free-fall, and the phenomenon of free-fall, will be the same on the train, moving or not. And if the window shades were down, you couldn't tell if the train was moving, at least not by doing experiments with gravity, experiments like dropping objects and measuring their acceleration.

There is an important generalization to be made from this. The results of any experiments in mechanics, and hence the laws that govern these experiments, are the same in all reference frames, at any place, moving or not, as long as the motion of the reference frame is steady, that is, at a constant velocity. Galileo pointed this out, and used it to argue that no experiment on the Earth could reveal whether the Earth was moving or not, since the outcome of the experiment would be the same either way. This general principle plays a fundamental role in both mechanics and the science of gravity. Einstein gave it a name, the **Principle of Relativity**. Note how misleading that name is. The principle says that the laws of physics are *not* relative to the reference frame; they are **invariant**, the same in all reference frames. Principle of Invariance would be a better name, but we are stuck; it's the Principle of Relativity.

Put the train in motion at some constant speed v, have someone on the train drop a stone, and consider what the trajectory of the falling stone looks like to someone standing on the ground. Assume they can see into the train car, and they can see the stone for the entire duration of the fall. The stone accelerates down at 9.8 m/s^2. It also has a constant horizontal speed v, as it stays with the train. The Principle of Relativity requires that the two components of motion, vertical and horizontal, are independent of each other. There is no horizontal force on the stone. It got its initial horizontal velocity from the train, but once let go, that force is gone. There is now only the force of gravity, straight down. The result of acceleration down plus constant speed horizontally is a curved trajectory. More precisely, it's a parabola. The arc gets steeper as the stone falls, since the stone is not just moving down, it's speeding up as it moves down.

Now do exactly the same experiment but without the train. A stone with some horizontal velocity is let go. That's the same as just throwing the stone horizontally. It is, in the language of mechanics, a projectile. The horizontal component of velocity continues unchanged, while the vertical component of velocity increases at 9.8 m/s^2. The trajectory is a parabola, as it was when the train was in the picture. The horizontal distance between the release of the stone, the throw, and its hitting the ground is called the range. This is shown in Figure 3.4.

In a reference frame moving horizontally with the projectile we would record the object's motion as being simply straight down, as if simply dropped. This is back on the train, dropping the stone. If instead of a tossed stone the projectile is a bullet shot horizontally from a gun, the kinematics are exactly the same. The trajectory is a parabola, and the greater horizontal velocity results in a longer range. If you could speed along-side the bullet, you would see it simply fall straight down to the ground, at 9.8 m/s^2.

The parabolic trajectory of a projectile curves down and eventually meets the ground. If the horizontal velocity is really fast, the horizontal distance, the range, is really long. We've been assuming the ground is flat, but in fact the Earth is round, and the surface curves gently down and away from the flight of the projectile. If the

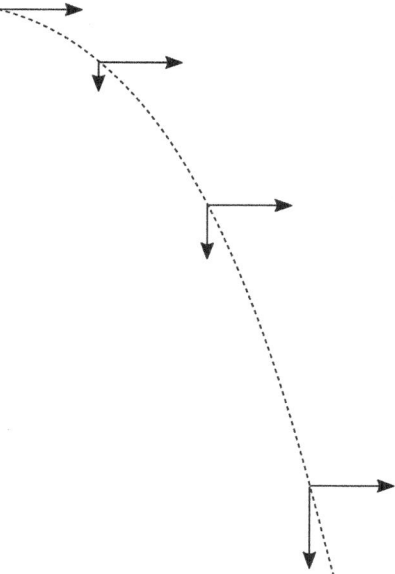

Figure 3.4. Several moments at equal time intervals during the flight of a horizontally thrown projectile. The horizontal component of velocity is the same at each moment, while the vertical component of velocity steadily increases. The resulting trajectory is a parabola.

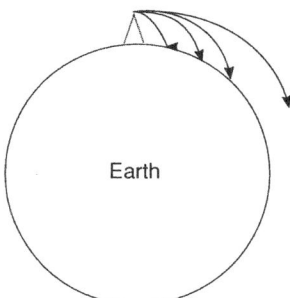

Figure 3.5. Projectiles thrown horizontally with progressively increasing velocity. With higher initial velocity, the range increases. At some point the initial horizontal velocity will be sufficient to put the object in orbit.

horizontal velocity is great enough, making the range long enough, the Earth will curve away just as fast as the trajectory curves down. The projectile continues to fall (at 9.8 m/s^2) but continues to miss the Earth. It will do this all the way around the Earth and the projectile is now a satellite; it's in orbit. Figure 3.5 shows some projectiles with successively higher velocities, until the one that finally makes it into orbit.

Just how fast would the projectile have to go to get into orbit? First figure out the range of a normal bullet, to see how close it comes. A typical muzzle velocity is 1000 m/s. If the bullet is fired horizontally, shoulder-high (approximately 2 meters off the ground), it will go about 600 meters before hitting the ground. There's not much curvature of the surface of the Earth in 600 meters, so the bullet won't even come close to making it into orbit. To calculate the speed it would take to achieve orbit may seem a complicated task, comparing the curvature of the trajectory and the curvature of the Earth. But there's an easy way to do it. Just use the orbit equation, Equation (3.7). What is the velocity required for an orbital radius equal to the radius of the Earth (plus the 2 meters shoulder-height)? This is a very idealized thought-experiment, assuming a perfectly spherical Earth with no obstructions over 2 meters high, but it gives us an idea of the speed required. When you do the math, the orbital speed is about 8000 m/s, eight times the speed of the bullet. Since the required orbital speed decreases for higher orbits – the r is in the denominator – real satellites have speeds somewhat less than this. They nonetheless need a significant boost, first the vertical to get to the orbital elevation, and then sideways to achieve the orbital speed.

As always, the mass of the object, the bullet or the satellite, never entered into the calculation. That's because it was pure kinematics, and with gravity, all masses respond the same. This fact is sufficiently important as to deserve a proper name. It's called the **Principle of Equivalence**, mentioned in Chapter 1. There are two versions of the principle at this point, and there will be one more as the science of gravity progresses. Later versions don't reject or revise the earlier; they add to them.

Once again it's Galileo who first clarified the principle. Heavy things and light things fall at the same rate. This is the first version of the Principle of Equivalence. The acceleration of gravity is independent of the mass, and even the weight, of the object. In modern terms, this means different dynamics – different weight, different force – do not cause different kinematics – acceleration. This is because the mass of the stone is on both sides of the equation, the gravitational force side, and the second law of motion side, the $\mathbf{F} = m\mathbf{a}$.

The mass of the stone will cancel, as long as the mass that is the coupling source of gravity is identical to the mass that is the inertial resistance to force. And this is the second way to express the Principle of Equivalence. This is Newton's formulation of the principle, that the gravitational mass m_g of an object, the property that causes the force of gravity, is equivalent to the inertial mass m_i, the property that resists the effect of any force, gravity included: $m_g = m_i$, so they cancel.

The equivalence of gravitational and inertial mass has the consequence of everything falling at the same acceleration, 9.8 m/s^2. It has the additional, related, consequence that the kinematic effects of gravity disappear in a reference frame that is falling along with the stone, falling at 9.8 m/s^2. If you do your physics in an elevator instead of a train, and allow the elevator to free-fall by cutting the cable, objects in this lab will not fall to the floor. The dropped stone will just float, stationary at the point of release. There will be no gravitational effects in the freely falling reference frame. The dynamics disappear with this kinematic change of frame.

The reverse is true as well, that is, an accelerating reference frame will result in gravitational effects, even if there are no massive objects around, no gravitational field. Do your physics in a spaceship, just drifting through space without the rockets burning, far from any gravitational sources. Nothing will fall in any direction, and the released stone will just float, exactly as it did in the falling elevator. But now fire the rockets so the spaceship is accelerating. All the loose objects, including the released stone, will fly toward the back of the ship, just as you are pressed back into the seat when a jet accelerates down the runway. In fact, all the loose objects will accelerate toward the back of the ship at the same rate, equal to the acceleration of the ship itself. The effect will exactly duplicate the effect of gravity, with the back of the ship as the new down. The dynamics appear with this kinematic change of reference frame.

This leads to a third version of the Principle of Equivalence. The effects of gravity are identical to the effects of an accelerating reference frame. This is the version that will eventually get us to the general theory of relativity.

We have identified two important principles in this chapter, the Principle of Equivalence and the Principle of Relativity. Both principles were recognized and used by Galileo. Both were named and strictly applied to the science of gravity by Einstein. Both play roles in the Newtonian theory of gravity, but it was Einstein's precision and enforcement of the principles that led to the downfall of the theory we have been working with in this chapter. We'll follow this principled undoing of the Newtonian theory, but first we should see what came before and what led to Newton's formula of the force of gravity.

4

Gravity Before Newton

The science of gravity began with Aristotle (384–322 BC). He explained why it is that what goes up must come down. The explanation is found in his physics. He also accounted for the motions of the planets. This is in his astronomy. And so we will start with Aristotelian physics and astronomy. The physics remained pretty much unchanged for millennia, but the astronomy was put through stages of revision, settling on the Ptolemaic model (Ptolemy, *c.* 110–*c.* 170). The plan is to clarify Aristotle's explanation for why a dropped object falls to the ground, and to clarify his methods in support of the explanation. And we will present the details of the Aristotelian–Ptolemaic description of planetary motion, also with an eye on the methods to derive the astronomical conclusions. This will be the historical entry to the science of gravity and the analysis of scientific method.

The modern science of gravity, as summarized in the previous chapter, applies the same laws of physics to phenomena on the Earth as to astronomical objects and events. It is, after all, the law of *universal* gravitation. Aristotle would have objected to this as an inappropriate and confused unification. At the time, and for a very long time after, the study of the physical world was separated into the terrestrial and the celestial. Physics was about what happens on the Earth, and astronomy described the heavens. Different laws applied. A department of astrophysics, as you might find at a modern university, would be viewed by Aristotelian and medieval scientists as a marriage of convenience, at best, or a huge blunder at worst. Doing experiments in an Earth-bound laboratory as a way to understand what happens on planets or stars is as misguided as experimenting on people to understand what happens in rocks. You might end up saying that the red rocks in the Grand Canyon get their color by blushing, embarrassed by all those tourists staring at them. People and rocks are importantly different kinds of things, understood by importantly different laws. The same is true, in the Aristotelian analysis, of terrestrial and celestial things.

The distinction between the terrestrial and celestial, and the corresponding separation of the sciences of physics and astronomy, were based on evidence. Aristotle noted that there is a uniformity observed in astronomical events that is missing here on Earth. Stars and planets move, but on unchanging, eternally repetitive trajectories. None of the lights in the sky ever goes out, and no new lights appear. On the Earth, by contrast, objects move in all directions, stop and go, and speed up and slow down. They can even be destroyed or created. In contrast to the perfect and eternal harmony in the sky, there is episodic imperfection on the Earth, and this difference demands two separate sciences.

The regularity of action in the celestial realm indicates a uniformity of composition. There is just one celestial element, aether, later called quintessence, the fifth element. There are four terrestrial elements, and everything on the Earth is composed of some combination of earth, air, fire, and water. Each of these elements has a natural motion, a way of moving when left alone, left to its own intrinsic nature. Aetherial objects, that is, everything in the sky beyond the orbit of the Moon, naturally moves in a circle, a perfect circle. This matches the perfect, recurring, eternal trajectories of the stars and the planets. The natural motion of terrestrial elements is in a straight line, up or down. The directions up and down are determined by the single point that is the center of the universe. Thus, earth, the element, naturally moves straight toward the center of the universe, while air moves naturally away. Water goes down, straight toward the center of the universe, and fire goes up.

This explains a lot. For example, the Greeks knew the Earth is round, and the Aristotelian system of elements and their natural motions explains why the Earth is round. The solid components of the Earth are composed primarily of the element earth. A collection of pieces all falling into a single point, in this case the center of the universe, would naturally form a sphere.

Notice how a subtle change in the wording of this account of the round shape of the Earth brings the Aristotelian theory very close to a modern version. Change the "pieces all falling into a single point" to "attracted to a single point" and this would not be out of place in a physics textbook today. Actually, both versions would fit. The original, the description in terms of falling, is the kinematic result of the second version, the dynamic cause. One key difference between Aristotle and our basic Newtonian version is that in the former the objects are falling toward a point in space rather than toward another physical object. It could be an empty point in space that determines the motion, the natural motion of objects.

The natural motions of elements gives Aristotle a way to explain and anticipate the motions of real objects on the Earth. What goes up, must come down. That's because what goes up is, in the usual experiment, composed mostly of earth, and going up is not only against its nature but also puts it above some air. The act of lifting or throwing a stone, displacing it from its natural place, is what Aristotle called violent or enforced motion. Once you let go, once the violence (the force) is stopped, the stone will follow its nature and return to where it belongs. Earth moves down, air moves up, and the stone is naturally located beneath the air, that is, closer to the center of the universe. The stone falls because it is in the nature of things made of earth, the primary element in a stone, to be below air.

When the stone is above air, it has the potential to fall, to seek out its proper place, and you can feel this potential if you are holding the stone. This is the stone's weight. Some stones have more potential than others, that is, they weigh more. You would expect the heavier object to fulfill its nature and seek its proper place with greater speed than does a lighter object, just as a more motivated person moves faster toward his or her goal than does someone who cares less. Expect a heavy thing to fall faster than a light thing. And this is exactly what usually happens. Drop a stone and a feather together and the stone falls faster. The heavy stone is composed almost entirely of earth, while the light feather must be made of some earth but a significant amount of air as well. This accounts for both the light weight of the feather and the slow descent of its fall.

There is agreement between theory and evidence in Aristotle's account of things falling. Unsupported objects fall, and the theory explains why. Heavy things are seen to fall faster than light, and the theory explains why. The agreement is not perfect, though, since Aristotle claims that the speed at which an object falls will be in proportion to its heaviness. This would have a 10 kg stone falling twice as fast as a 5 kg stone, and that just doesn't happen.

The disagreement in the details between theory and evidence reveals a more systemic difference between Aristotelian and modern physics. We would call it a flaw in his method. Aristotelian physics is strictly qualitative; there is no mathematics and no quantitative measurement. Had he applied mathematics to physics, Aristotle might have been less satisfied with the theory of falling objects. But there is a reason Aristotle did not mix mathematics with physics, and consequently did not pursue quantitative analysis in either the theory or the evidence. Mathematics, primarily geometry in Aristotle's time, is about perfection and precision. Life on Earth is a mess, an ongoing whirlwind of disruption, displacement, and disorder. Geometry applies to, for example, perfect circles, not wobbly misshapen circles as we can draw or find in the real terrestrial world. The lines of a triangle are exactly straight, but no line in nature is exactly straight, since natural motion is generally interrupted and corrupted by some violent intervention. Mathematics cannot describe what is imperfect and distorted, so mathematics cannot be applied to physics.

Mathematics, geometry in particular, can be applied to the uniformly circular motion of the planets and stars. Astronomy, in other words, can be done quite differently from physics. Astronomy can take full advantage of the best mathematics, since celestial objects and their motions are, as is appropriate to mathematics, uniform and perfect. This shows that the transition from Aristotle's physics, in which he explains why things fall to the ground, to Aristotle's astronomy, in which he describes the orbits of celestial objects, is abrupt. It is somewhat anachronistic, even a little bit Whiggish, for us to look for a unifying theme, namely gravity, between the two, but much of our modern theory of gravity was developed as revision and reaction to Aristotle, both the physics and the astronomy, so we need to understand both. From the turmoil that is the Earth, then, we'll look up to the harmony that is heaven.

There are seven planets in the Aristotelian cosmological model, five that we regard as planets today – Mercury, Venus, Mars, Jupiter, and Saturn – and the Sun and the Moon. Each of these wanders through the unchanging backdrop of the stars. Each follows essentially the same kind of orbit and is described in the same way. So, to understand the orbit of one is to basically understand them all. This allows us to talk in terms of the generic planetary orbit.

Like the stars on the celestial sphere, a planet orbits the center of the universe once per day. The shape of the orbit is perfectly circular. This is the diurnal rotation of celestial objects, from east to west, and it accounts for the visible motion that is the rising and setting and movement across the sky. A planet has an additional motion, on its own spherical shell and at its own pace, very slowly from west to east. This is why the planet appears to move through the pattern of stars, usually showing up a little bit east, compared to where it was the day or night before. The Sun, for example, moves eastward through the constellations of the Zodiac at roughly one degree per day. This brings it once completely around in the course of a year. The path of the Sun as it moves against the background of the stars is called the ecliptic.

The Roman architect and scientist Vitruvius (85–20 BC) provided a homely analogy of the Aristotelian planetary system. "If seven ants were to be placed on a potter's wheel, and as many channels were to be made around the center of the wheel, growing in size from the smallest to the outermost, and the ants were forced to make a circuit in these channels, then as the wheel was spun in the opposite direction . . ." (Vitruvius, 2001, p. 111) the motion of the ants would resemble that of the planets.

Notice that there is no mention yet of the Earth. This is astronomy, not physics. The planets orbit the center of the universe, as the ants circle the center of the wheel. As it happens, the Earth occupies that point at the center of the universe, but it plays no causal role. The Earth has only an incidental role in describing the motion of the planets.

Figure 4.1 shows the basics of an orbit of any of the seven planets. The planet is at point P; the Earth is at E. The entire model rotates clockwise once every 24 hours, as does the celestial sphere of the stars. This is the diurnal rotation. The individual

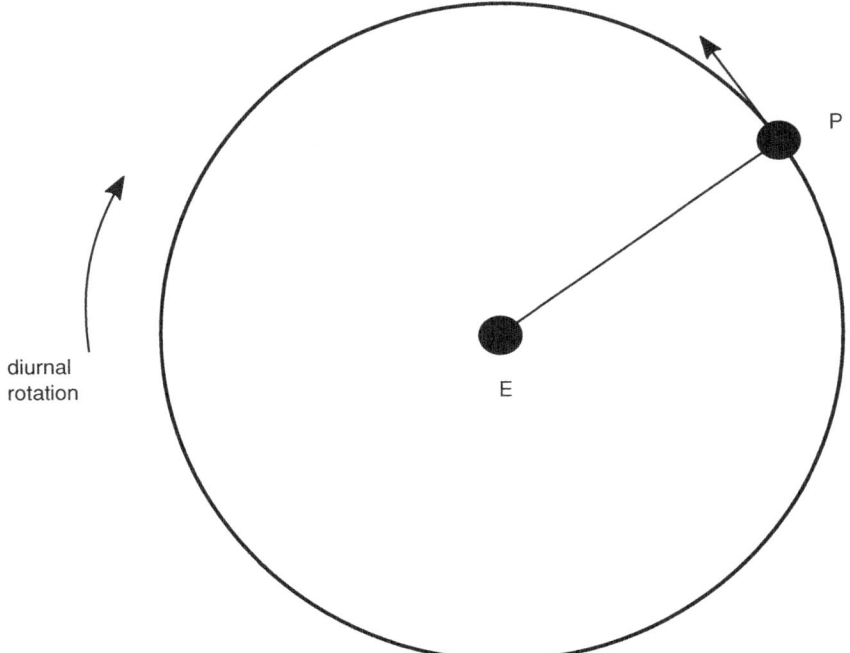

Figure 4.1. The orbit of a planet in the most basic Aristotelian model. The Earth E is at the center of the universe. All the celestial objects rotate clockwise around the center, once around in 24 hours. This is the diurnal rotation. Each planet P also rotates slowly counterclockwise. This results in its moving a little bit each night eastward against the background of the stars.

orbit of the planet is much slower and depends on the particular planet. Mars, for example, takes more than a year to orbit counterclockwise.

The basic Aristotelian account of planetary motion is a paradigm of simplicity and elegance. It is composed of perfect concentric spheres and eternally uniform motion. But it suffers significant empirical inadequacy, noted even at the time. The planets other than the Sun and the Moon are seen to get brighter and dimmer from time to time, suggesting that they get closer and further from the Earth. This defies the model of perfect circular orbits centered, even coincidentally, on the Earth. Furthermore, the motion of these planets as measured against the background of the stars is not uniformly west to east. The planet is seen to stop and change direction, only to soon resume its west-to-east motion. This retrograde motion defies the model of constant angular speed. With these anomalies, we might then consider the Aristotelian model to be only a first-order approximation of the nature of planetary orbits, with higher-order terms to follow.

The first significant modification of the basic Aristotelian model was made by Apollonius of Perga (mid third century to early second century BC), and

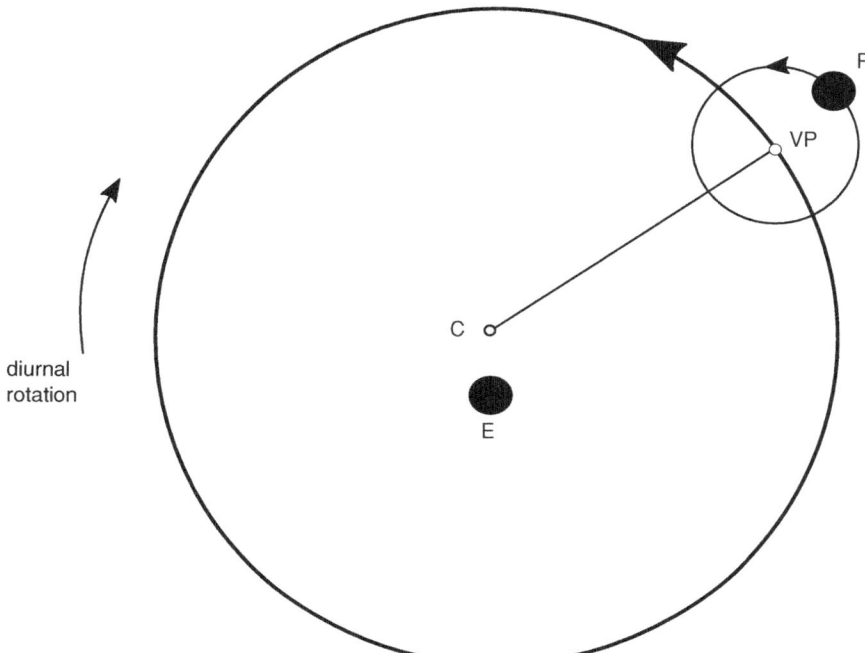

Figure 4.2. The planetary model of Apollonius and Hipparchus. All celestial objects have the 24-hour diurnal rotation clockwise. A planet P also orbits on a circular epicycle. The center of the epicycle itself moves in a circular orbit, the deferent, centered at the point C. The center of the epicycle is marked VP to indicate that it is a void point, an empty point in space that plays an essential role in determining the motion of the planet. The Earth E is not quite at the center of the deferent. The displacement between C and E is the eccentric.

Hipparchus (second century BC). In the model of Apollonius and Hipparchus, each planetary orbit requires two circles, an **epicycle** and a **deferent**. The planet itself orbits on the epicycle, a circle that is centered on a point that also moves on another circle, the deferent. This is shown in Figure 4.2.

Note again the clockwise diurnal rotation of the entire model, once around in 24 hours. Motion on the epicycle and deferent are there to account for the movement of the planet against the background of the stars. The epicycle results in both retrograde motion of the planet and its changing brightness. Retrograde happens when the planet is closest to the Earth and its orbit on the epicycle is in the opposite direction from the motion on the deferent. The size and period of the epicycle can be adjusted to match the observed movement of each individual planet. To achieve the retrograde motion, the radius and period of the epicycle must result in the planet moving backward, from the perspective of the Earth, and hence crossing its own orbital path. This produces a **rosette**, a loop-de-loop pattern. The path of the planet is no longer a perfect circle, but it is composed of two perfect circles. Thus is the harmony and uniformity of celestial motion preserved. It is naturally eternal and repetitive and without deviations.

The center of the epicycle, that is, the center of the planet's actual circular motion, is an empty point in space. Julian Barbour calls these **void points** (Barbour, 2001, p. 175). A void point is a point in space where no object, no matter, is located but that plays an essential role in describing the motion of a planet or even an active role in causing features of the motion. It is worth keeping track of the void points in early astronomical models, because they seem entirely contrary to our modern sense of what can hold a planet in orbit. It takes a real thing with mass to provide the centripetal force needed by Newtonian theory. But in this ancient account of orbit, the planet seems to be tethered to nothing at all.

The Aristotelian model with its Apollonius–Hipparchus revision clearly contains a void point. In fact, the new model has two void points. If you look closely at Figure 4.2 you will see that the Earth is not exactly at the center of the deferent circle. This displacement between Earth and orbital center is called the **eccentric**. It is a feature of the Apollonius–Hipparchus model to accommodate observations that the planets spend a longer time on one half of their orbit than on the other. Even in the original Aristotelian model it was the point in space, the center of the universe, and not the object, the Earth, that determined the center of a planetary orbit. The Earth just happened to be at that point. The eccentric has the effect of exposing the implicit void point in the Aristotelian model, showing that the motion of a planet is not in relation to another object but to a feature of space.

With the epicycle, deferent, and eccentric, the revised model of planetary orbit has lost much of its original simplicity and elegance. It's the price paid to keep the model empirically accurate and true to the theoretical requirement that the motions are, or at least are composed of, perfect circles. This compromise highlights a three-way balance that is generally the challenge of scientific method. The empirical data, in this case how the planets move and change brightness, cannot be ignored and there is an ongoing obligation to observe more, and more precisely. A second constraint is theoretical. The description of nature must be consistent with the generally accepted theoretical understanding of the fundamentals. In this case, the planetary orbits must be composed of perfect circles rotating at uniform speed. The empirical and theoretical must fit together, although the fit is never perfect, and there is an obligation to improve the fit in response to new and better data. There is a third consideration in evaluating scientific results, the more abstract and sometimes aesthetic virtues of a theory such as simplicity, beauty, or elegance. We can call these simply conceptual virtues. None of these three factors has ultimate authority over the others. At this point in the history of planetary astronomy, the conceptual virtues

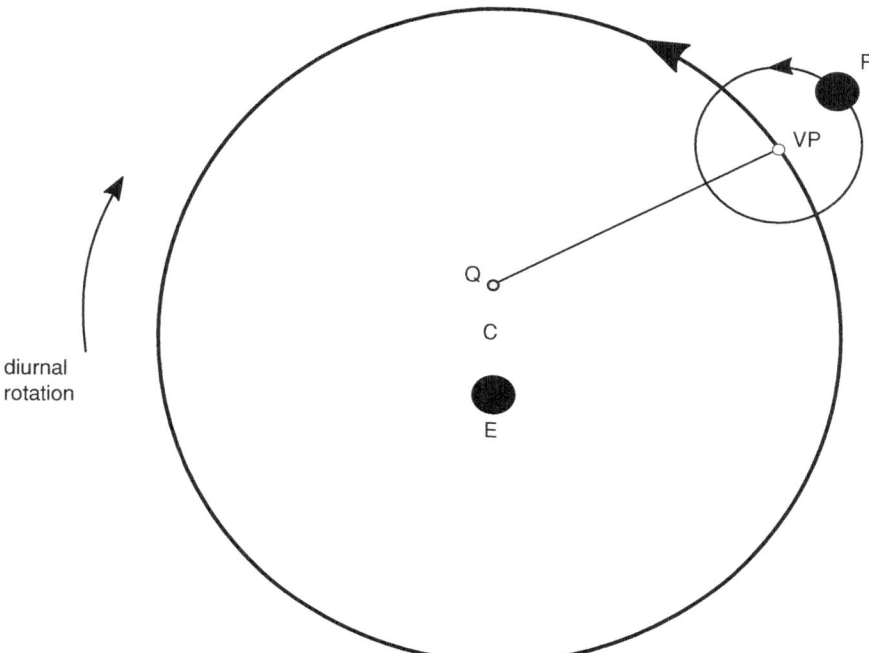

Figure 4.3. The planetary model of Ptolemy. The diurnal motion, epicycle, deferent, and eccentric are all the same as in the Apollonius–Hipparchus model of Figure 4.2. Ptolemy adds the equant at Q. The line between Q and the planet P rotates with a constant angular speed.

have been sacrificed to accommodate the empirical and theoretical. It won't always be like this.

Commitment to the Aristotelian theoretical requirement of perfect circles, and continued improvement in measuring the motions of planets, led to further complication of the model. Ptolemy in the first century AD added one more term of correction to the basic Aristotelian model, the **equant**. He kept the Apollonius–Hipparchus features of epicycle, deferent, and eccentric, and thereby retained the rosette shape of the planetary orbit and the void point at the center of the epicycle. Only the void point at the center of the deferent got some revision.

The Ptolemaic model still has the deferent as a perfect circle centered at a point a bit removed from the Earth, but the angular speed of the deferent void point, the center of the epicycle, is not constant around this point; it is constant around the equant point that is opposite the Earth and an equal distance from the center. This equant is another void point. In Figure 4.3, again the planet is at P and it orbits on an epicycle. The center of the epicycle, a void point, orbits on the deferent that is geometrically centered at C, the eccentric point. The line in the figure is drawn not to the center of the deferent but to Q, the equant. This is the line that rotates at a constant angular speed.

The theoretical purpose of the equant is to accommodate more precise data on the speed of the planet across the sky. The planet moves faster on one side of the circle than it does on the other. Rather than have the angular speed of orbit vary, the point of determining the angular speed is relocated to the equant. This is consistent with the Aristotelian theoretical ideal of perfection in celestial phenomena. The orbits are perfect circles and the rate of rotation is perfectly uniform.

The Ptolemaic model has a lot of adjustable parameters. The radii of both the epicycle and the deferent, and the period of orbit around each, can all be adjusted to match each planet in its movement against the background of the stars and the timing of its retrograde motion. The theory prescribes the basic concepts and components of the model, perfect circles and uniform rotation, but then the phenomena themselves dictate the details.

It may seem to us that the theoretical insistence on perfect circles in any model of planetary motion has resulted in some implausible results. The epicycles, for example, orbits upon orbits, may seem *unnecessarily* complicated. Simplicity in a theory sometimes has to be compromised, but this may be a reckless forfeiture. And the void points, empty points in space that determine the planet's orbit, seem to require that a bit of nothing does some real work.

These days, it is usually the epicycles that draw our criticism of previous models of planetary motion. These extra orbits violate our code of simplicity and elegance, and so should be avoided if possible. But a quick look at the Solar system as currently described shows that we really should have no principled objection to epicycles. The orbit of any moon is on an epicycle around the planet, as the planet orbits around the Sun on the deferent. This is wholly unobjectionable because the center of the epicycle orbit that tracks around on the deferent is occupied by an object, the planet. And since the planet has mass, it plays a dynamic role. As the source of a gravitational force, the planet acts as a tether for the orbiting moon. So, it's not the epicycle *per se* that would be objectionable; it's an unoccupied orbital center that presents the dynamic impossibility. The problem is the void point.

The void point at the center of the epicycle, certainly, and at least one of the two void points at or near the center of the deferent, probably, are dynamic in the sense of having a causal role in determining the trajectory of the planet. From our perspective this is problematic, since it leaves a fundamental aspect of the cause, and hence the explanation, of the orbit to rest on nothing at all. But in the Ptolemaic context, the void points are acceptable. Aristotelian and Ptolemaic astronomers were not interested in the causes of planetary motion. Since the motion was circular it was natural and didn't need a cause. For this reason, astronomy was a strictly kinematic endeavor, with the goal of describing planetary orbits but not explaining them. As simply descriptive tools, that is, in a strictly kinematic role, the void points present no conceptual challenges. They are not causal, that is, dynamic, and consequently not necessarily asserted as real features of nature.

The last part of the devaluation of the void points, not claiming them to be real physical things in nature, is compatible with the usual interpretation of Ptolemy as an instrumentalist. **Instrumentalism** is an attitude toward science that regards utility for calculation and prediction as the only important virtue of a theory. Truth is irrelevant. If a theory works it's a good theory. End of story. If the model of planetary orbit is *just* a useful model, or, as Ptolemy is translated, a hypothesis, then there is no need to worry about the physical possibility or the causal mechanism. It's just a way to keep track of where and when a planet will appear. But there are reasons to think that Ptolemy was not strictly instrumentalist about his account of the planets, and, more importantly, that subsequent astronomers evaluated the model not simply in terms of its pragmatic virtues like accuracy of prediction but in terms of its being true or false. Ptolemy inherited from Aristotle the theoretical idea that the planets were carried along in their orbits by being embedded in transparent, crystalline, spheres. It's the spheres that rotate. And in all his work, Ptolemy saw the need to adjust the radii of deferents so that the spheres that carried the planets did not intersect, as they can't since they are solid. If the spheres were merely calculating devices, ways of our thinking, it wouldn't matter if they intersected. Only a realist interpretation of at least this aspect of the model would necessitate the adjustment.

Ptolemy the person, and his attitude regarding the physical reality, or not, of the components of the planetary model are not as important as the status of the Ptolemaic model itself. Later interpretations of the model, and reactions to it at the time of Copernicus, were decidedly in realist terms. Georg Peurbach (1423–1461) presented the Ptolemaic model in *Theoricae novae planetarum* (1454) as physical reality, complete with solid quintessential spheres to carry the planets. This was the model to which Copernicus responded. And even after Copernicus, Georg Rheticus (1514–1574) argued against the possibility of the equant as "a relation that nature abhors." (Quoted in Hoskin and Gingerich, 1999, p. 87.) The criterion for rejecting the idea, in other words, is not that *we* abhor the equant, as we would for pragmatic reasons, but that *nature* abhors the equant. It's about what is or is not in nature, not what works or does not work for describing nature.

These interpretations, historical and modern, regard the void points in the Ptolemaic model and the rosettes it retains from Apollonius and Hipparchus as real features of nature. So are the quintessential spheres, and this takes the dynamic burden off the void points. With the solid sphere holding the planet and rolling it around on a circular orbit, the center of the sphere, the location of the void point, does not have to act as a guide or tether. In this way, the void points in the Ptolemaic model, even a physically real Ptolemaic model, have only a kinematic role, allowing Ptolemy and Ptolemaic astronomers to accept them as neither puzzling nor problematic.

Instrumentalist or not, the Aristotelian–Ptolemaic astronomy was not concerned with the causes of celestial events. The goal was simply a model of movements that matches the observations. The method was to do whatever it takes, as was said at the time, to save the phenomena. Using the theoretical tools of the day, not forces and fields but spheres and quintessence, one adjusted the parameters in whatever way it took to agree with past observations and predict where planets would be seen in the future.

The method was quite different in physics, the science of terrestrial events. In this case, knowing about causes, and hence being able to explain why things happened as they did, was important. Why is the Earth round? Because the element earth naturally moves toward the single point that is the center of the universe. Why does a heavy object fall to the ground faster than a light one? Because the heavy one contains more earth and consequently has more potential to seek its proper place, closer to the center of the universe. The faster fall is what you would expect, given the composition of the object.

In physics, the explanation starts with a logical deduction. Aristotle's scientific method is often described, and often with some disapproval, as being a kind of top-down reasoning. It starts with general principles, like the natural motion of elements, and reasons *a priori*, without further evidence, as to what nature must be like. The logic of Aristotelian physics is deductive and the results are both certain and necessary. The Earth *must* be round, and heavy things *must* fall faster than light. These are not simply generalizations about features of nature; they are laws of nature.

How does this aspect of Aristotle's method compare to modern science? Some things are the same, for example the idea of laws of nature. What goes up, *must* come down, by the *law* of gravity. But some things are different, for example the requirement that the conclusions be drawn with absolute certainty. Aristotle regarded certainty as a necessary requirement of knowledge. There was no doubt, no uncertainty, in Aristotelian science. By modern standards this is dogmatic and contrary to the open mind expected of a scientist. A requirement of scientific method, or at least scientific attitude, is a recognition of the uncertainty of any result and a willingness to consider alternatives and challenges.

Aristotelian physics wasn't all *a priori* logic. There was an important role for observation. Heavy things do in fact hit the ground before light. This is directly observable. There is evidence that the Earth is round, and so its shape is indirectly observable. During a lunar eclipse, the shadow of the Earth on the Moon is rounded not straight. And ships sailing away from harbor are seen to disappear bottom first, then the mast, as if they are sinking or descending along the curved surface of the Earth. There is even evidence in Aristotle's physics that shows the Earth cannot be moving. It is a very indirect observation that the Earth stands still, but it is an important contribution to the science. Drop a stone from the top of a tall tower and observe that the stone falls straight down. If the Earth were moving, the stone would fall some distance away, as the Earth and tower moved out from under it. Galileo will have a lot to say about this argument, but for now it shows one of many examples of Aristotle's *a posteriori* reasoning. It shows the role of evidence in his scientific method.

No scientific method is an algorithmic sequence of steps, but Aristotle's approach to understanding terrestrial nature generally has observations coming before theorizing. We see what needs to be explained. A theory is then deduced. And since the logic of deduction is foolproof, with no possibility of error or uncertainty, there is no need to follow with empirical testing. This is another big difference between Aristotle's method and what happens now in science. The Aristotelian theory is not tested, and there are no controlled experiments in which the scientist sets up specific circumstances and regulates specific variables. And, with no quantitative observations or analysis, because perfect mathematics can't describe the imperfect events on Earth, there is no precise comparison between theory and data. Theory claims not only that a heavy thing falls faster than a light but falls faster in proportion to its heaviness. This seems never to have been measured, since measurement is an application of mathematics to nature, and since there should be no need to confirm a deduced theory.

Aristotle's scientific method puts observation in what philosophers of science now call the **context of discovery**. This is the formative stage of scientific theorizing, when ideas are suggested but not endorsed. It must be followed, if the process is to be genuinely scientific, with a **context of justification**. This means doing whatever it takes to test the ideas, to challenge those that fail and provisionally endorse those that pass. For Aristotle, the proof of a theory was in its deductive derivation. This results in certainty. Modern scientific method, and this will be one of Galileo's important contributions, will require not just the reasoning and evidence that lead to a hypothesis, but also after-the-hypothesis evidence. And because the testing has an essential empirical component it will be unavoidably uncertain. No experiment is perfect, and no experimental results are beyond question. It will be Galileo's achievement to show how to apply perfect mathematics to the imperfect phenomena on Earth. He will bring mathematics to physics and allow the marriage of astronomy, already mathematical, and physics.

Aristotle and Ptolemy provided an account of why things fall to the ground and how planets orbit, and in this sense they initiated the science of gravity. Their account differs in many ways from what is currently on the books in the chapter on gravity, but the modern theory was formed largely in reaction to the ancient theory, so understanding the latter will help in understanding the former. The same is true for the methods used. That is, the Aristotelian scientific method differs in some important ways from what we require of scientific method today. The modern method will make more sense, and the reason for various components will be more important, by comparison.

5

Early Modern Astronomy

Astronomy, we now know, is an important part of the science of gravity. But the importance is recognized only within the conceptual framework of a *universal* force of gravity with the same laws at work in the heavens as on the Earth. There was no such connection in the science of Aristotle and Ptolemy and through the medieval period. For these scientists, there would be no reason to consider a link between things falling to the ground and planets orbiting whatever they orbit. In astronomy, after all, what goes up stays up. And in the heavens, what moves continues to move in a uniform, uninterrupted way. There is no reason to ask what causes the motion, since it is eternal. In early astronomy, unlike physics, there was no role for dynamics, and that's where gravity would show up.

To see where the modern concepts of gravity came from, we will have to trace the unification of physics and astronomy, that is, the physical sciences of phenomena on the Earth and in the sky, and the unification of kinematics and dynamics. It's when the Earth itself moves, and when it joins the other planets in the heavens, that both aspects of unification become necessary. It starts with Copernicus.

Nicholas Copernicus (1473–1543) worked exclusively on astronomy. He had nothing to say about why things on the Earth fall to the ground. In that sense, his contributions are not directly relevant to the science of gravity. And as an astronomer in the tradition of Aristotle, Copernicus made no effort to explain why the planets move as they do. His accomplishment was a kinematic revolution with no regard for dynamics and hence no opportunity or need to theorize about gravity. It's only with historical hindsight, looking back at what happened next, that we recognize Copernicus as a step toward the model of the Solar system Newton worked with in developing the concept of universal gravitation. Copernicus set up a Solar system to be revised by Kepler and held together by Newton. Copernicus and Kepler clarified the kinematics for which Newtonian dynamics provided the force.

In the Preface to his *Principia Mathematica* Newton described the road to universal gravitation this way. "All the difficulty of philosophy seems to consist in this – from the phenomena of motions to investigate the forces of nature, and then from these forces to demonstrate the other phenomena." (Newton, 1995, p. 4.) His term "philosophy" here means natural philosophy, the discipline we would call natural science. And the challenge he describes, the "difficulty," is met with a summary of scientific method as applied to physics. From a precise description of motion, the task is to figure out what causes the motion. From the kinematics is inferred the dynamics. From the motion is inferred the force. And then, knowing the force one can predict other phenomena. This is how the dynamic theory can be tested. Newton's method must start with a clear description of the phenomena. That's where Copernicus comes in.

The understanding of planetary motion was more or less unchanged from the basics Aristotelian–Ptolemaic system until the Copernican revolution. With separate epicycles, deferents, and equants for each individual planet, there were enough adjustable parameters to respond to new data and save the phenomena. Copernicus kept a lot from Aristotle. His planetary orbits were composed of perfect circles, and the rotation of each circle was perfectly uniform. The heavenly spheres in *On the Revolutions of the Heavenly Spheres*, published in 1543, are not the individual planets but the invisible quintessential spheres that hold each planet in its orbit. Altogether, Copernicus used almost all of the pieces of the Aristotelian model; he just put them together differently.

Copernicus raised no objection against the epicycles or eccentricity of the Ptolemaic model. Like Rheticus, he found the equant to be the principle flaw. And each planet was guided by a different equant, so there was no one point in the system that determined the constancy of orbital motion. In the Preface and Dedication to Pope Paul III in *On the Revolutions of the Heavenly Spheres*, Copernicus faults advocates of the equant model as having "admitted a great deal which seems to contradict first principles of regularity of movement. Moreover, they have not been able to discover or to infer the chief point of all, i.e., the form of the world and the certain commensurability of its parts." (Copernicus, 1952, p. 507.) It's not so much that the Ptolemaic model has the center of the system in the wrong place but that it has no single, unifying center at all.

By putting the Sun in the center of the planetary system, Copernicus hoped to restore the fundamentals of Aristotelian uniform motion. The basics of the Copernican model are familiar, and their presentation in Figure 5.1 is perhaps unnecessary.

The Sun is at the center of the planetary orbits, and the Earth, now just one of (at the time) six known planets, orbits the Sun in 365 days and rotates on its axis in 24 hours. The Earth moves. This explains the retrograde motion of the other planets without resorting to the complicated loop-de-loop pattern in their orbit, the

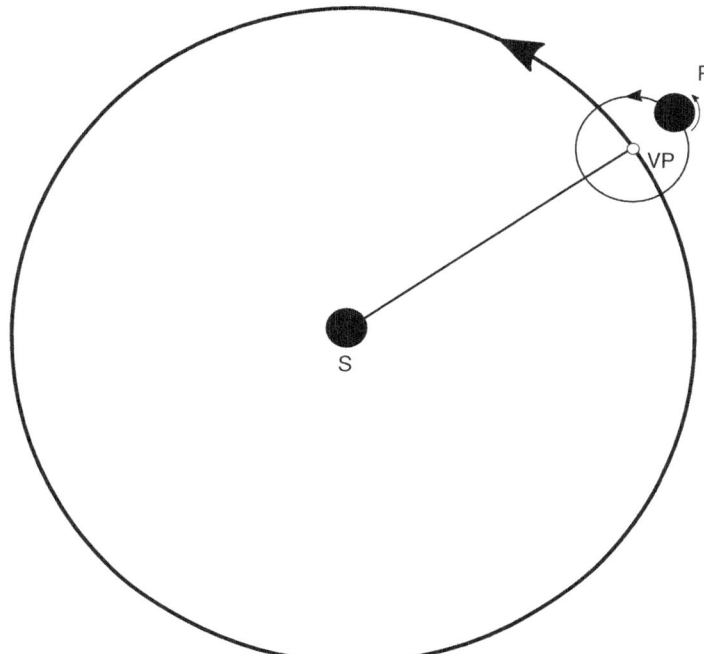

Figure 5.1. The basic Copernican model of planetary motion. The Sun S is at the center of the system of planets. Each planet P, including the Earth, orbits the Sun, S. The Earth also rotates on its own axis, once every 24 hours. Planets are still on epicycles that orbit void points VP on the deferent centered on the Sun. These are small epicycles, and the planet never crosses its own path.

rosettes. A planet such as Mars appears to move backward when the Earth passes by on its own orbit. That's the retrograde motion.

But these are just the basics, literally just the first chapter in the book. Deeper into the Copernican mathematics we find that each planet is still on an epicycle, a perfect circle centered on a moving void point on a deferent. There is an important difference between this and the Ptolemaic model, in that the radius of the Copernican epicycle is smaller and the period of epicycle orbit is long enough that the planet never loops back across its orbital trajectory. As shown in Figure 5.2, the period and radius of the epicycle can be adjusted to make the orbit eccentric, but it does not result in a rosette. In this case, the epicycle rotates in the same direction as the deferent but at twice the rate. These parameters can also be fine-tuned to produce an elongated orbit that approximates an ellipse.

Using the terminology from Thomas Kuhn (Kuhn, 1957, p. 67), a "major epicycle," as in the Ptolemaic model, produces a qualitative change in the apparent motion of the planet, that is, the stopping and reversing of retrograde motion. A

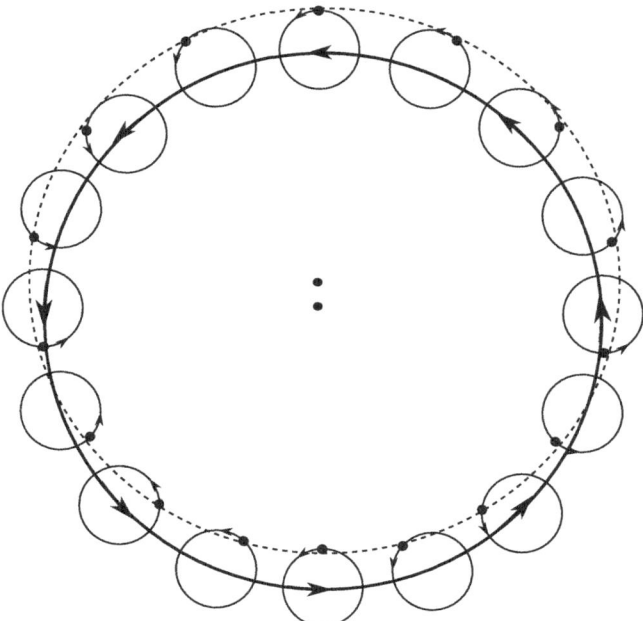

Figure 5.2. A minor epicycle that rotates in the same direction as the deferent and at twice the rate. The resulting path of the planet is shifted, as shown by the dashed line. The center of the deferent is no longer the center of the orbit, so the orbit is eccentric.

"minor epicycle," as in the Copernican model, produces only quantitative changes, speeding up and slowing down. A major epicycle results in a rosette; a minor epicycle does not.

There are more complicating details in the Copernican model, more fine-tuning required to match the data. The deferent of the planetary orbit is not centered exactly on the Sun. It is centered on a point, a void point, that is on an epicycle that is itself centered on yet another point, another void point. This ultimately central point, point X in Figure 5.3, is the same point for all the planets, and it is determined relative to the Sun and the Earth. When you look closely, the Copernican model is no simpler than the Ptolemaic model. There is quite a mess in the middle, and a lot of empty points in space doing a lot of work to determine the orbit of each planet.

The so-called heliocentric model of the universe doesn't really have the Sun at the center. But then the geocentric model didn't have the Earth at the center of the orbits either. Both models were put together by the Aristotelian guidelines of perfect circles and uniform rotational speeds. And both models end up rich in void points.

The responsibilities of the void points can be lightened if we remember that astronomy at the time was interested in the description of motion but uninterested

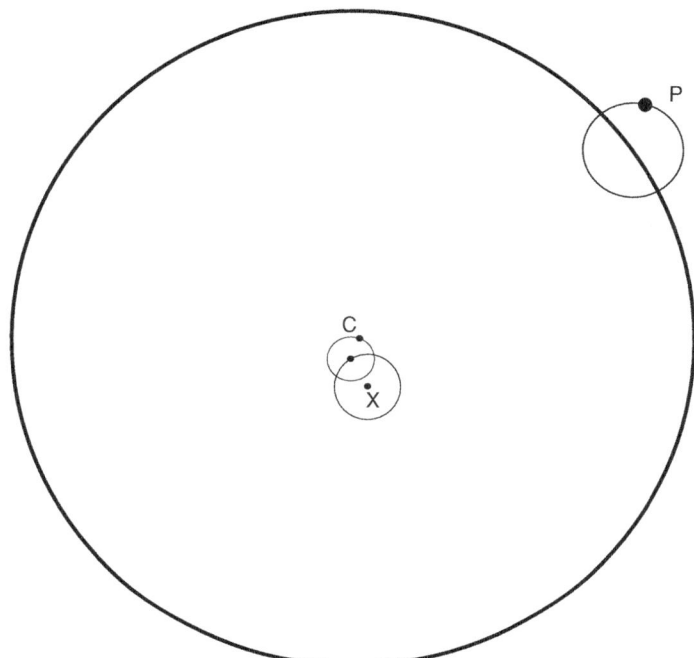

Figure 5.3. The complication at the center of the Copernican model of planetary orbit. The planet P is on a minor epicycle that orbits a deferent centered on point C. C is itself on an epicycle that orbits on a deferent centered on point X. X is the same point for all the planets, thus there is a single center of the planetary system, but it is not exactly at the Sun. X is a void point, as is C. X is determined relative to the Sun and the Earth.

in the cause. Those empty points in space play no dynamic role. They weren't expected to hold the planet in orbit or to be a tether point always tugging the planet into its circular trajectory. Copernicus still had the planets embedded in heavenly spheres, with the implicit understanding that it is the rigid spheres that move and hold the planets in place.

The void points are even less troublesome, and there is nothing controversial or condemnable about putting the Earth in motion, if the entire astronomical account is regarded as unreal, simply a useful calculating tool to account for past positions of the planets and to predict future positions. This is instrumentalism, to value the model as exactly that, as *just* a model, with no presumption that the details are real features of nature. On this interpretation, there is no claim that the Earth is really moving; it's just that if we do the calculations *as if* the Earth orbits the Sun, we get accurate results with no need of an equant.

The instrumentalist interpretation of the Copernican model would avoid the obvious challenges, both empirical and scriptural, to the claim that the Earth moves.

Observations, both our own day-to-day sensations of stability and, as we will see, precise astronomical measurements, indicate a stationary Earth, a *terra firma*. So does the Bible. A last-minute and unauthorized (and anonymous) Introduction to *On the Revolutions of the Heavenly Spheres* advocated this pragmatic understanding of the new cosmology. It was a colleague of Copernicus, Andreas Osiander, who saw to the final production of the book as the author lay on his deathbed, and added the Introduction.

The instrumentalist interpretation, however, would not have motivated the need for gravity. Only real motions require forces. With the Sun at, or at least near, the center of things, and with the Earth in eccentric motion around the Sun, the Copernican model is getting close to the elliptical orbits described by Johanes Kepler (1571–1630). These are the "phenomena of motions" from which Newton discovered the forces. The Earth and other planets must really be moving, and moving as the model indicates, for there to be the need for a real cause and real forces. The science of gravity, or at least the history of how the science developed, calls for a realist interpretation of the astronomy.

The development of planetary astronomy from Copernicus to Kepler will require a dramatic concession to the ideals of Aristotelian astronomy. Heavenly motion will lose the perfection of circles and uniform angular speed. But there is a compromise model, one that retains the circles and avoids the implausible notion of a moving Earth. This is the Tychonic model.

Tycho Brahe (1546–1601) is notable for his state-of-the art astronomical observations, and for his opposition to the Copernican model of planetary orbits. He provided the most precise data available before the first use of telescopes in 1609. Using these data and the theoretical requirements of circular orbits, Tycho, as he is known, achieved the same empirical success as Copernicus while keeping the Earth stationary. It's not at all difficult. The Tychonic model has all the planets orbiting the Sun. In this way it is exactly like Copernicus, and it includes the minor epicycles both on the planet's deferent and at the deferent's center. The only real difference is that unlike the Copernican model, the Earth is not one of the planets, and Tycho has the Sun orbit the Earth. So, the Earth stands still – no offence to sensation or scripture there – while the Sun orbits the Earth and the planets orbit the Sun. These basics, ignoring the minor epicycles, are shown in Figure 5.4.

At first, the Tychonic model looks contrived, a desperate attempt to hold off the obvious conclusion that the Earth is just one of the planets. It keeps the Earth from moving, but at the expense of a thousand inconveniences. But that judgment really depends on how you look at it, that is, which point you choose as your reference, the origin of your coordinate system. The Tychonic system is exactly the same as the Copernican, just drawn from the perspective of the Earth rather than the Sun. They differ only in this choice of reference frame.

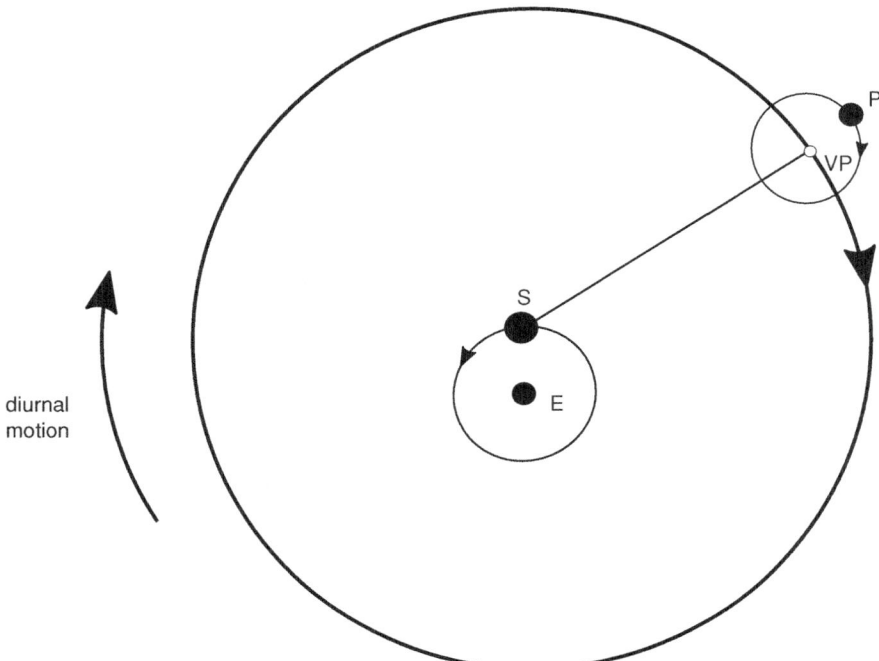

Figure 5.4. The basic Tychonic model of planetary motion. The Sun S orbits the stationary Earth E. The planet P orbits on an epicycle, centered on the void point VP, as the void point orbits on a deferent that is centered on the Sun. As with the Aristotelian–Ptolemaic models, the entire system rotates clockwise, once every 24 hours. This diagram is for a superior planet such as Mars, Saturn, or Jupiter. An inferior planet, Mercury or Venus, would have a deferent radius that is less than the distance between the Earth and the Sun.

In fact, if we ignore the minor epicycles, there is no fundamental difference in any of the three planetary systems we have described so far. The Tychonic hybrid helps to demonstrate this. Ptolemaic epicycles and deferents can be adjusted to exactly match the Tychonic system, and from there it is just a change of reference system to get the Copernican system. It's worth showing how this works, step by step.

First, consider the idea of relative motion. Think about two ships A and B going in opposite directions and passing each other in the open ocean. From the deck of A one would say that B is moving. From the deck of B, A is moving. We can determine which ship is really moving, or if both are really moving, by looking at the water and listening for the engines. But now take away the water and the engines. That is, consider two ships, or any other objects, drifting through otherwise empty space. Inertia alone carries them along, and there is no medium through which to move. From the perspective of A, B is moving. From the perspective of

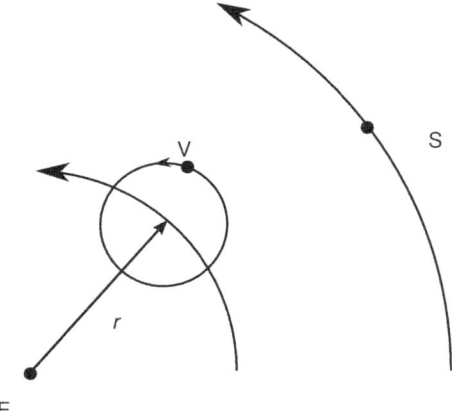

Figure 5.5. The basic Ptolemaic model of the orbit of Venus. The radius *r* of the deferent is adjustable to match the observed motion of Venus. The entire system has a diurnal rotation, moving clockwise, once every 24 hours.

B, A is moving. There is not only no way to tell which is really moving, there is no way to define which is really moving. Moving or not is not determined, in either sense of determined, except by reference to some other object. Moving or not is a matter of reference frame, and that is a matter of choice.

Apply this analogy to the objects in space, the Earth and the Sun. There is no medium to move through – it is genuinely empty space – and no engines propelling either object. Thus there is no way to determine which one is moving. One moves relative to the other, but it is a matter of choosing a reference frame to determine which is the one and which is the other. The Sun moves with respect to the Earth; the Earth moves with respect to the Sun.

Now use this concept of relative motion to demonstrate the kinematic equivalence of the Copernican and Tychonic systems, and with a little allowable adjustment, the Ptolemaic system. Start with Ptolemy, and draw the basic model of the orbit of an inferior planet, that is, a planet that orbits between the Earth and the Sun. Venus, for example, is an inferior planet. Its deferent, as shown in Figure 5.5, is inside the deferent of the Sun. The planet itself is on an epicycle, moving on the deferent. The radii of both deferent and epicycle, and the periods of orbit of both deferent and epicycle, are all adjusted to match the observed astronomical data.

Simply adjust the radius of the planetary deferent to be the distance between the Earth and the Sun. That is, make the planet's deferent the same as the Sun's, and put the Sun at the center of the planet's epicycle. All of these adjustments are allowed within the Ptolemaic model. They are even encouraged, since the spirit is to do whatever it takes to save the phenomena, that is, to comply with the evidence. This fine-tuning of the planet's deferent eliminates the obvious void point at the center of

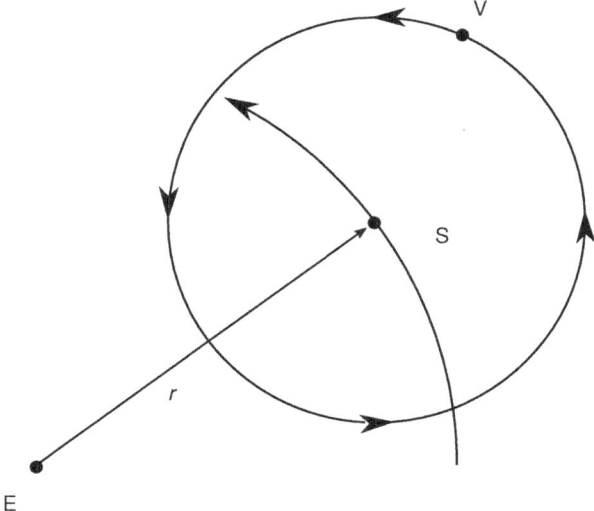

Figure 5.6. The basic Ptolemaic model of the orbit of Venus, with the radius of the deferent adjusted to equal the distance between the Earth and the Sun. The Sun is moved to the center of the epicycle, so Venus now orbits the Sun rather than a void point. The Sun orbits the Earth. The result is exactly the Tychonic model of the orbit of Venus.

the epicycle, since that point is now occupied by the Sun. It also turns the Ptolemaic system into the Tychonic. Figure 5.6 shows the results of the adjustments; it is also the basic Tychonic system.

Now look at the Sun S and the Earth E in Figure 5.6, and remember the facts about relative motion. From the perspective of E, S is moving. From the perspective of S, E is moving. Draw the situation from the perspective of S by applying the opposite motion to E. This results in Figure 5.7, and it is exactly the basic planetary motion of the Copernican system. The Sun is the center of the orbits of both Venus and the Earth, with Venus orbiting between the Sun and the Earth.

So, does the Earth move or doesn't it? And don't forget that people lost their lives or their personal freedom over this issue. If we consider only the geometry of the planetary orbits, that is, only the description of the motions, the kinematics, there is no real difference between the Tychonic system, in which the Earth does not move, and the Copernican, in which it does. It just depends on where one chooses to locate the origin of the reference frame. It's a matter of perspective. The kinematics are equivalent, and since at the time astronomy was only about kinematics, there was no astronomical difference. Only when the questions are asked about the cause of the motion and the dynamic stability of the system will real physical differences separate Tycho from Copernicus.

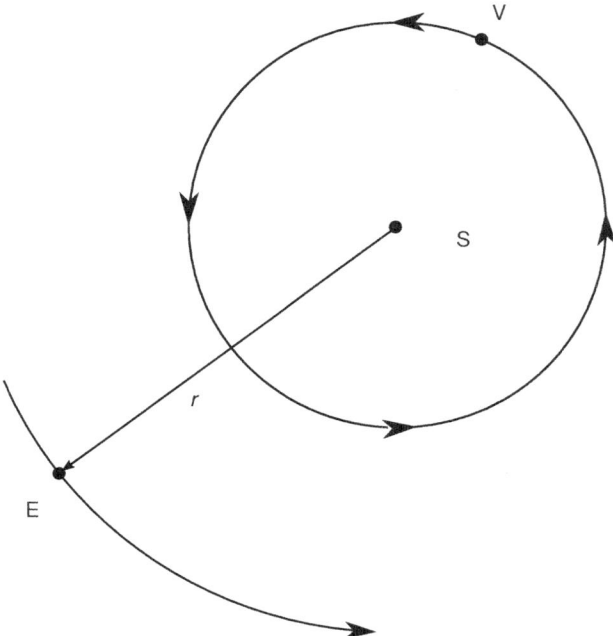

Figure 5.7. The same physical situation as in Figure 5.6, but from the perspective of the Sun rather than the Earth. The relative motion between the Sun and the Earth can be set in either the reference frame of the Earth-at-rest or the reference frame of the Sun-at-rest. Choosing the Sun at rest results in the basic Copernican model of planetary motion. This shows that the only difference between the Copernican system and the Tychonic (and, with allowable adjustments, the Ptolemaic) is the choice of perspective.

Despite this lack of kinematic distinction, Tycho did contribute an observational discovery relevant to the science of gravity as highlighted in terms of void points. He observed a comet in 1577 with such precise measurement of its distance as to confirm that it was celestial. Comets had been observed before this, but it was most commonly believed that they were atmospheric phenomena, hence not so far from the Earth as to threaten the eternal harmony of the heavens. But Tycho measured the comet to be at planetary distance, and with a trajectory that intersected planetary orbits. This would be impossible if the planets were carried by real, solid, quintessential spheres. But, if the spheres are not really there, then what is it that holds the planet in its circular orbit? What guides the planet to circle the moving point on the deferent? These are obvious questions and concerns from our perspective, since we require the kinematic description to have, or at least allow, a dynamic mechanism. We put mathematical and physical astronomy together. But these concerns were not so explicitly combined until Kepler. When the dynamic question became important, the void point was seen to play an unlikely role as the tether of the orbit. It is an empty point in space that directs the planet's circular orbit. This will make the role of the void point more important, and potentially more problematic.

It is exactly the dynamic void points in models of planetary orbits that drew Kepler's attention and objection. His reasoning derived from his more mystical understanding of nature.

Kepler thought of a planet as having both a mind and a soul. The mind directs the trajectory of the planet and thereby determines the orbit. The mind needs a real object, not a vacant point in space, as a guide to center the orbit. It can't be a vacant point between the Sun and the Earth, the "mean sun" as in the Copernican model; it has to be the real Sun. In Kepler's own words, "A mathematical point, whether or not it is the centre of the world, can neither affect the motion of heavy bodies nor act as an object toward which they tend." (Kepler, 1992, p. 54.)

Kepler was explicitly interested in the cause of planetary motion, not simply an accurate description. He published his model in 1609, in a book whose title is usually abbreviated as *Astronomia nova*, New Astronomy. But the full title reveals his interest in physical causes of planetary motion, *Astronomia nova* *ΑΙΤΙΟΛΟΓΗΤΟΣ, eu Physica Coelestis, tradita commentariis de Motibus Stellae Martis*. The one Greek word amongst the Latin is "aetiological," meaning the study of causes. This emphasis, in which the motion of a planet must make physical sense and have a physical cause, makes the dynamic void points implausible.

The result, the Keplerian model of planetary orbit, has no void points whatsoever. The price is dear, at least from the perspective of Aristotelians. Kepler abandons the restriction to perfect circles in describing orbits. Careful observation, and a wild abandon from the theoretical restriction, led to Kepler's first law: The orbit of each planet is an ellipse with the Sun at one focus. The Sun is exactly at one focus of the ellipse, focus being the Latin word for hearth. Thus, "The sun is the fireplace of the world." (Kepler, 1952, p. 855.) The hearth is occupied by an object, and the object is the physical guide for planetary orbit.

This achieves the effect of the eccentric without using a single epicycle, major or minor. In this sense, the new system is simple and elegant. In the Aristotelian sense, the new system is a mess, a squashed and elongated insult to celestial perfection. Here is one theoretical worldview contrasted with another, and it's not objectively obvious whose aesthetic criteria should rule the scientific decision.

The Keplerian model is shown in Figure 5.8, but with the eccentricity greatly exaggerated.

An ellipse has two foci, and only one of them is occupied in the Keplerian planetary model. The other is void. But, the unoccupied focus has no role in either describing or directing the orbit of a planet. It's not really part of the model. As it turns out, an observer at the unoccupied focus would observe a nearly constant

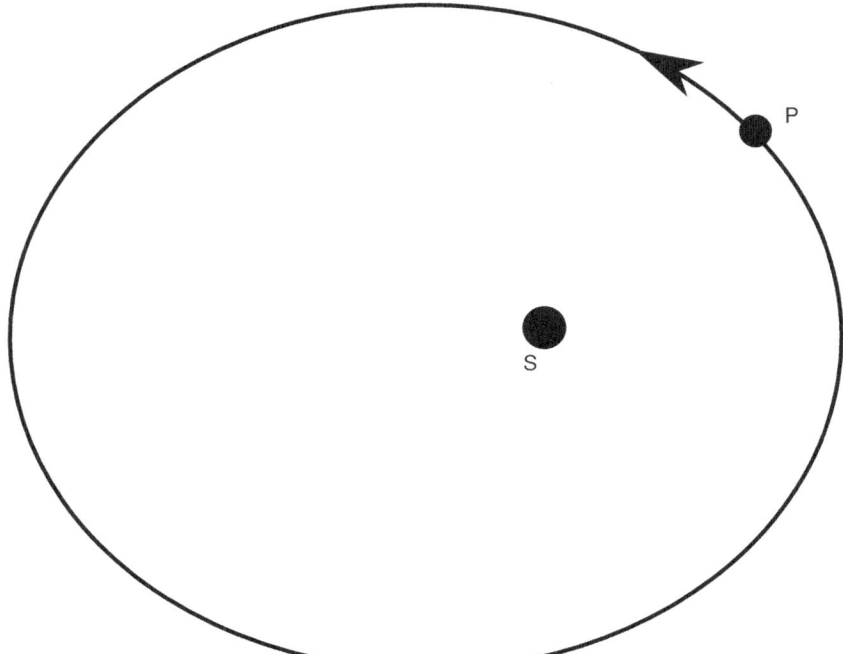

Figure 5.8. Kepler's model of planetary motion. The Sun S is at one focus of the elliptical orbit. The eccentricity of the orbit is greatly exaggerated in this picture. There are no epicycles, deferents, or void points.

angular speed of the planet, and in this way the point is roughly equivalent to the Ptolemaic equant. This may afford some retrospective credit to Ptolemy for antici- pating this feature, but the unoccupied focus is not an active ingredient in Kepler's model.

Kepler simply accepted the fact that the planet did not orbit the focus of its ellipse with a constant angular speed. The celestial motion is neither circular nor uniform. But there is a simple relation between the two imperfections in that the planet speeds up when it is closer to the Sun. A pattern emerges from the data, and this leads to Kepler's second law: The line from the Sun to the planet sweeps out an equal area in equal time. Over a short period of time, this line pivots around the Sun to form a shape like the slice of a pie. At the far reaches of the ellipse the slice is long but narrow, since the line pivots slowly. Close to the Sun the slice is short but fatter, since the line pivots more quickly and sweeps out a more generous serving. Kepler showed that the area of the long skinny slice is always equal to the area of the short fat slice.

At the time there was neither an explanation for this relation nor a precise math- ematical expression for the area of the slice. It's not a triangle, since one side is curved, like a pie crust. The only way to calculate the area is to consider very small sections, slices formed over very short times, and regard the curve as approximately straight. Then it's a triangle and the formula for the area is known, and the small increments can be added up. We can see this as the preface to calculus.

In what sense are these laws? They are clearly generalizations derived from empirical data. But they are neither explained nor explanatory. Since the regularities emerge from observation rather than theoretical or mathematical derivation, they provide no understanding of why the orbits are elliptical or why the area is uniformly drawn. These laws are simply a summary of the facts, not an explanation of the facts. Nor are the summaries much help in explaining other astronomical phenomena. It would be nothing more than saying that's just the way things are. For example, why does Mars speed up when it gets closer to the Sun? Because it sweeps out equal areas in equal times. That doesn't advance our understanding of the motion of the planet since it only relates one aspect of the observation to another. There is no suggestion of the underlying mechanism or the cause of the motion that would necessitate the speeding up or the equal areas. Explanation requires this kind of necessity in the connections, and this is missing from the empirical generalizations that are Kepler's laws.

There is a third law, added in 1619, ten years after the first two. It is the most mathematically precise. Kepler's third law: The ratio of the square of the period of orbit to the cube of the semi-major axis of the orbit is the same for all planets. This is not at all obvious and it couldn't be a coincidence. The need to explain this remarkable feature of all the planetary orbits will lead to Newton's universal gravitation.

For clarification, the major axis of the orbit is the line drawn through both foci of the ellipse and extended to the ellipse itself. It's the longest diameter. The semi-major axis is simply half this distance. It is also the average distance between the planet and the Sun. For the Earth, that distance is 1.5×10^8 km. This gets its own unit, the Astronomical Unit, AU. So the average distance between the Earth and Sun, the semi-major axis of the orbit, is 1 AU. The period of the Earth's orbit is one year. In these units, AU and years, it's easy to calculate the ratio of the square of the period P^2 to the cube of the semi-major axis R^3. It's $1^2/1^3 = 1$. Kepler's third law says that using these same units, the ratio will be 1 for every planet. Mars, for example, has a period $P = 1.88$ y, and an average distance to the Sun $R = 1.52$ AU. Do the math and $P^2/R^3 = 1$. That's the law.

By eliminating the perfectly circular orbits, Kepler brought on the new challenge of providing a driving mechanism to move the planets. Circular orbits, as simply the uniform eternal motion of celestial bodies, required no cause. It was natural. But elliptical orbits and the changing speed of the planet called for a dynamic drive. It was Kepler's concern for dynamics, the physical cause of the planetary motion, that led him to reject the void points in previous models. It motivated him to apply physics to astronomy. His physics was inventive, but the dynamic explanations he gave of planetary motion were short-lived. The Sun, according to Kepler, emits a sweeping, magnetic sort of force that both holds the planets in elliptical orbit and pushes them around. The Sun is both tether and propulsion, at the universal hearth, holding the system together and keeping it in motion.

The magnetic action of the Sun results in elliptical orbits, but not closed orbits. That is, the planet does not return to the same point after one complete orbit. The major axis of the ellipse precesses. The **precession** was known to Kepler, but he offered only a sketch of an explanation in terms of differential action of the different poles of the magnetic Sun and planet. The kinematic consequences are clear, and were at the time. There are no loops or petals in any one orbit of a planet, but over the repetition of many orbits, a rosette-like pattern is produced.

But the void points were gone. No empty point in space, and in fact no feature of empty space, has any causal power. This is progress when viewed from what Einstein called **Mach's Principle**. Named for Viennese physicist Ernst Mach (1838–1916), and cited by Einstein as inspiration for his general theory of relativity, Mach's Principle insists that all spatial and temporal properties of things and events are determined in relation to other physical things. There is no reference to space itself as a thing, absolute space. All motion, for example, is relative motion, relative to objects. And all forces are caused by physical objects, not by features of space. You can see how void points, both in their kinematic role of locating an orbit, and their dynamic role of tethering an orbit, violate Mach's Principle.

Mach's Principle is inspired by a strict allegiance to empiricism, that no scientific idea be without direct empirical foundations. There are no empirical foundations for the idea of absolute space, that is, space as an entity with its own intrinsic properties. All we observe is position or motion relative to other objects. By empiricist Machian standards, talk of motion relative to an unobserved, and unobservable, point in space violates the most basic requirement of science, direct evidence. It's no better than mysticism. So, good riddance void points.

Notice how Mach's empiricism connects the two senses of determining a property. Since we cannot determine motion except by reference to some other thing, there is no meaning, no reality of motion except by reference to some physical thing. Since we cannot observe space or a point in space like a void point, there is no sense in referring a property like speed or orbital trajectory to space or a point in space.

There are always trade-offs in science and, in the case of Kepler, removing the void points came at the price of adding another mystical aspect of planetary orbits. There is nothing between the Sun and a planet, no medium, no connective strings or poles, and yet somehow the Sun exerts an influence on the planet to keep it moving and hold it in orbit. Galileo found this mysterious attraction between the Sun and planets unacceptable. The more modern reference is action at a distance, a causal interaction between two distant objects with no contact and nothing between them. As we saw in Chapter 2, the modern way of dealing with action at a distance is in terms of a force field that fills the space surrounding both objects. There was no field concept at the time of Kepler, and so his suggestion of action at a distance was regarded with suspicion and, in the case of Galileo, outright rejection.

Kepler deserves credit for his early attempt to bring dynamics to the understanding of the planets. He got the kinematics right, with the three laws, but he got the dynamics wrong. He set the table for Newton, both by providing the kinematic model of planetary motion in the succinct form of the three laws, and by setting the methodological example of merging physics and astronomy.

The history of the science of gravity at this point shows two parallel lines of inquiry now coming together. Aristotle contributed to, arguably even initiated, both physics and astronomy, but he kept the two sciences entirely separate. Things fall to the ground; that's physics. Planets orbit in the sky; that's astronomy. We see them as demonstrations of the same phenomenon, gravity, but Aristotle and Aristotelians saw them as fundamentally different. Physics cannot be done mathematically, since the perfect figures of geometry do not match the sloppy shapes in real life on the real Earth. Mathematics applies without error or approximation to astronomical orbits, as they are perfect circles. And the uniform motion around the perfect celestial circles is eternal and unchanging, so there is no need for a cause. It is natural motion. Thus, there is no application of dynamics to astronomy; it is pure kinematics. Motion on the Earth, by contrast, comes and goes and changes in many ways, and it is worth asking why. That's dynamics, and it has a role in physics. This is what kept physics and astronomy distinct for nearly two millennia, from Aristotle to Kepler.

The first hint that physics and astronomy might belong in the same department came when Copernicus put the Earth in motion and, more importantly, put it into the same realm as astronomical objects like planets. The loss of its distinctive position in nature suggests that the Earth doesn't need its own distinctive science. The explicit merger of physics and astronomy, and hence of kinematics and dynamics, was begun by Kepler. Galileo furthered the cause. Newton tied the knot.

6

Connecting Physics and Astronomy

One of the notable methodological accomplishments of the Copernican revolution was to note that scientists cannot trust their immediate sensations to report what nature is really like. Things are not as they appear. What looks like the Sun rising is really the eastern horizon descending. The Earth doesn't feel like it's moving, but it is.

Our most basic observations, in other words, are deceptive, and a fundamental task of scientific method is to interpret what we observe by using other things we know. Our understanding of nature helps us make proper sense of our experiences. To put it in explicitly scientific terms, theories influence observations.

Galileo (1564–1642) explicitly recognized this necessary balance between observation as the basis for theory and theory as a guide to observation. This was one of his most important contributions to scientific method. On the Second Day of the *Dialogue Concerning the Two Chief World Systems*, Simplicio, the advocate of the old, geocentric world system, insists that "the senses and experience should be our guide in philosophizing." (Galileo, 1953, p. 248.) But Salviati, the spokesman for the new, heliocentric world system, cautions that "first impressions . . . may easily deceive us," so we must "put aside the appearance, on which we all agree, and to use the power of reason." (Galileo, 1953, p. 256.) Galileo's methodological achievement was to show not only how to merge astronomy and physics, but how to connect theory and observation in a way that gets the most out of the collaboration.

Galileo contributed no new or revised model of planetary astronomy. Rather, he adopted and advocated the Copernican model exactly as it's found in the first chapter of *On the Revolutions of the Heavenly Spheres*. He seemed to have ignored the messy details of the minor epicycles and the contrived position of the center of planetary deferents, the so-called mean Sun. His interest, and the world system of the dialogue that brought on trouble with the Inquisition, had the Sun at the center of things and the Earth moving in two ways, once around the Sun in a year and once around its own axis in a day. He also seemed to have dismissed Kepler's suggestion of elliptical orbits. Galileo was generally put off by Kepler's mystical explanations of interactions between distant objects, and perhaps more fundamentally reluctant to give up the Aristotelian requirement that planetary orbits be perfect circles. This remaining Aristotelian influence will play a role in Galileo's physics as well as the astronomy. It will show up in the new concept of **inertia**, the inherent property of an object to continue in motion. For Galileo it will be inertia to continue moving in a circle rather than, as we now say, in a straight line. Galileo introduced the ideas that led to our understanding of inertia, and that was an important step toward the modern science of gravity, but it came with some Aristotelian baggage. There was a scientific revolution in progress, but it did not happen all at once.

There was no Galilean model of planetary orbits, nor was there any Galilean contribution to the precise data on the motions of the planets. That is, he offered no quantitative improvement to astronomical knowledge. Given his disregard for the minor epicycles of Copernicus and the ellipses of Kepler, empirical precision does not seem to have been a priority for Galileo, at least not in astronomy.

But of course, there was the telescope, and this was his qualitative contribution to astronomical evidence. The mountains on the Moon, the phases of Venus, and the moons of Jupiter all served as circumstantial evidence in support of the Copernican model, the new world system. Thus, the sort of non-mathematical analysis that had been the hallmark of physics, the science of terrestrial phenomena, Galileo brought to astronomy. This is astrophysics, but perhaps not in the way we had hoped to celebrate.

Galileo's most substantive contribution to the science of gravity was his work on the kinematics of free-fall. He brought the precision of mathematics down to Earth. By figuring out how to manipulate and measure the positions and speeds of projectiles, he managed to apply the uncompromising perfection of mathematics to the fast and, in some cases, unnatural motion of falling stones and cannon balls. This is the revolutionary methodological synthesis of astronomy and physics that set up the modern science of universal gravitation. It helped to unify the understanding of motion *on* the Earth and motion *of* the Earth.

Start with the motion of the Earth. To advocate the Copernican model of planetary motion is to deny the obvious, the facts that the Earth *feels* stationary and the stars and planets *look* like they are moving across the sky. But denying the obvious is routine science. Things are not as they appear, and that's exactly why we need science. At the time of Copernicus, and Galileo, the claim that the Earth moves defied not just brute appearances but careful measurement and coherent theoretical reasoning as well. It had a lot going against it, a lot of science. Two long-standing arguments against the idea that the Earth moves needed to be explicitly confronted. One was based on astronomical evidence, the other on terrestrial, thus covering both evidential bases.

Tycho explicitly made the astronomical case against the moving Earth. His reasoning, that is, his interpretation of the evidence, was based on the well-known phenomenon of **parallax**. Get a visual bearing on some distant object, a tree in the park, say, or just a chair across the room. Move to a different position and note that the bearing, the angular position of the tree or the chair, is different from this new perspective. This is parallax, the change in angular position of an object that results from your own change of perspective, your own movement. You can even do this while sitting down by changing perspective from left eye to right. Close one eye and look at your thumb held at arm's length. Change to the other eye, back and forth, and your thumb appears to shift back and forth. That's parallax.

Tycho reasoned that if the Earth moves then the angular position of distant stars will change as our observational perspective changes. This is stellar parallax, and measuring angular positions of astronomical objects was Tycho's strength. He could do it with cutting-edge precision. Note that the motion of the Earth that Tycho was testing is the yearly orbit around the Sun, not the daily rotation on the Earth axis, and the change in position should be considerable if the two measurements are made six months apart, from one side of the Sun to the other. With due diligence, Tycho looked for evidence of stellar parallax but found none. The Copernican theory made a false prediction, and on this basis Tycho rejected the theory and its core claim that the Earth moves. At least in this regard, things *are* as they seem.

There are three noteworthy details in this test of the Copernican theory.

First note the logical relation between theory and evidence, as it is characteristic of scientific testing. The hypothesis, the Copernican theory, is tested by making an observable prediction, stellar parallax. More to the logical point, the prediction is a consequence of the hypothesis. If the Earth moves then stellar parallax will be observed. If the hypothesis is true then a particular phenomenon will be observed. This is the logical core of empirical testing in science, and hence one of the fundamental links between theory and evidence. It's a theme we will see repeated, with variations and increasing detail, as the scientific method develops. Tycho has the scientific method exactly right.

Next note Tycho's attitude, or, since we can't and shouldn't try to read his mind, his approach to testing the hypothesis. He is an advocate of the old world system and he is convinced that the Earth does *not* move. He nonetheless entertains the Copernican hypothesis as if it's true. The reasoning is not, the Earth is stationary and let's prove it. Rather, let's suppose the Earth moves and try to disprove it.

Both the logic and the attitude match what is often called falsification in science, and this is the third key aspect in Tycho's testing. The method is not to prove a theory but to disprove a theory. The logic is ideally suited for falsification, proving a hypothesis to be false. If the hypothesis makes a prediction that turns out to be false, then the hypothesis, it seems, must be false. Hasty accounts of scientific method often leave it at that, suggesting that falsification is decisive. One false prediction and you're out. But clearly, and it is clear from the case of stellar parallax, it can't be that simple. The Copernican theory made a prediction that, using the very best observational equipment, turned out to be false. But the Copernican theory is true. A true theory can make a false prediction. We need to understand specifically how this happened in the case of testing the Copernican theory by stellar parallax, and more generally why falsification is both important in science but also inconclusive.

Parallax depends on distance. You can eye-to-eye parallax your thumb, but if you try to do it for a distant tree or a more distant mountain top, or an even more distant star, you won't be able to detect the change in angular position. It's there, but too small to measure. There are two distances that factor into the parallax calculation, the distance to the object and the distance between the two perspectives, the distance you move from one observation to the other. If you know one of these, you can use the measured change of angle in the parallax to calculate the other. Astronomers, now knowing the diameter of the Earth's orbit around the Sun, use parallax to measure the distance to relatively nearby astronomical objects. But a prediction of the angle of parallax, as at the time of Tycho, will depend on knowing both distances. For Tycho to know that the stellar parallax would be measurable by his tools, he had to know both how far away the stars are and how far the Earth would move according to the Copernican theory.

By Tycho's estimate, the celestial sphere of stars was not much beyond the most distant planet Saturn. Nature, by Aristotelian account, does not accommodate a vacuum, so one celestial sphere must be contiguous to another. By this calculation, the stars would be close enough, and the hypothesized orbit of the Earth large enough, that stellar parallax would be detectable using Tycho's tools. The parallax was not detected, so the hypothesis was rejected. We see the problem. He had the stars too close. But at the time, and science always has to be done at the time, the theoretical and empirical pieces fit together to conclude against the idea that the Earth moves. Advocates of the Copernican model had to live with or explain away the challenging evidence. They had to make the universe bigger, putting the stars simply too far away for the parallax to be measurable. It wasn't just a little bigger; it had to be at least 700 times bigger than previously thought.

The important methodological point here is that a hypothesis that makes a false prediction can be saved by blaming some other theoretical idea that was part of the testing process. There was credible theoretical reason to think that the measurements would be good enough to detect the effect, but those theoretical reasons were challenged. This makes testing, even in the case of falsification, always open to challenge. And don't think that the matter of parallax was finally settled with the use of a telescope. Remember that Galileo added no precise quantitative data to astronomy, even with his telescope. Stellar parallax was not detected until 1838. The Copernican theory had to live with the unfulfilled prediction of parallax for almost three centuries. And of course when parallax was finally observed, the challenge was met but the theory was not confirmed. Making a true prediction doesn't prove that a theory is true any more than making a false prediction proves the theory is false.

The parallax evidence against the motion of the Earth came from observing things in the sky. There was also evidence from observing things on the Earth. Both astronomy and physics tell against the Copernican model. The terrestrial evidence again comes from an argument by Aristotle, but this time the test of the theory focuses on the alleged diurnal motion of the Earth, the rotation once around the axis in 24 hours.

Aristotle mentioned the experiment only briefly, but Galileo made a very big deal of it in the *Dialogue Concerning the Two Chief World Systems*. His strategy was to show not only that Aristotle failed in his attempt to disprove the claim that Earth moves, but that the attempt was based on a fundamental mistake in the Aristotelian physics. So much gets done, both scientifically and methodologically, in Galileo's treatment of Aristotle's argument that historians have given this a proper name. It's called the tower argument.

Aristotle's argument was this. If the Earth is moving, then an object, a stone, say, dropped from the top of a tall tower will be seen to fall some distance away from the base of the tower. This is because in the time it takes for the stone to fall to the ground, the Earth and the tower will have moved some distance to the east. The stone will fall to the west of the base of the tower. But when the experiment is done, the stone is seen to fall straight down. It falls to the base of the tower. So, the Earth is not moving.

Note the structure of the argument. The logic is identical to Tycho's parallax argument. It is again the standard approach to falsification. And again, the (as we know) true hypothesis has made a false prediction. Falsification is not decisive. And as with the parallax example, it is worth pinpointing exactly why the false prediction does not unequivocally disprove the hypothesis. The way to do this is to ask what Aristotle was assuming when he made the prediction, and whether there was good reason at the time to make that assumption.

In the parallax case the assumption was about the distance to the stars, and it was motivated by the established theories about empty space and celestial spheres. In the tower argument the assumption is that the horizontal motion of the stone will abruptly stop as soon as it is released. It moves horizontally with the Earth, on the hypothesis that the Earth is moving, only as long as someone is holding it and pulling it along. Left to itself it falls straight down. This is because the natural motion of a stone, being composed mostly of the element earth, is straight down. And when left alone the stone has only its natural motion to follow. It has no horizontal natural motion, and to get it to move horizontally requires an outside influence. That's violent motion. Only celestial objects naturally move horizontally. So, there was at the time this good reason to expect the stone to stop moving horizontally and fall some distance away from the base of the tower. Peer review at the time would have endorsed this prediction.

Galileo confronted the tower argument by challenging the underlying theoretical idea that the stone's horizontal motion would stop when it was released. This required reaching deeply into Aristotelian physics and denying one of its most fundamental laws about natural motion. Galileo argued that the stone can have two natural motions, one vertical and one horizontal, so that when released it will continue along with the moving tower and fall exactly below the point of its release. Unlike parallax, it's not that the predicted effect would be too small to detect. In this case there would be no effect at all.

The demonstration of horizontal natural motion, what would later be called inertia, was not by direct empirical means but by an **analogy** and ultimately a thought experiment. The real difficulty in determining whether the Earth moves or not is that we are stuck on the thing itself. If we could step off and look back, we could simply see from that external perspective whether the Earth moves. So it would be useful to consider an analogous system for which we do have the detached perspective. Drop a stone from the top of the mast of a ship and note whether it falls some distance behind when the ship is moving forward. We can tell from shore if the ship is in fact moving, and from this we can see if on-the-ship experiments can indicate the motion.

Doing the experiment on the ship would show that the stone falls straight down even when the system moves, and thereby challenge the authority of the tower argument. But at the time, authority, in particular the authority of Aristotle, seemed to be more important than evidence. Galileo had to change this methodological priority as well as the theoretical claim about natural motion. He says in the dialogue, "our discourse must relate to the sensible world and not to a world on paper." (Galileo, 1953, p. 113.) He was not alone in this reaction against the attention to the authoritative world on paper. Peter Severinus (1542–1602), a chemist and medical scientist, was blunt, advising those who study science to "burn up your books . . . buy yourself stout shoes, travel to the mountains . . . " (Quoted in Eisenstein, 1979, p. 472.) And Francis Bacon (1561–1626), a clear spokesman for a bottom-up evidential method of natural science, makes clear that, "the discovery of things is to be taken from the light of nature, not recovered from the shadows of antiquity." (Quoted in Dear, 2009, p. 58.)

This sounds right, that science is not based on what you find in books but what you find in nature. You should believe not what you're told but what you see. Conclusions are not from authority but from reason and evidence. All of this sounds right, but it's a significant overstatement and a bit naïve.

The role of authority that Galileo opposed in scholastic science is still present to a degree in science today. It has to be. Some reliance on and respect for authority is not only inevitable in science, it is necessary and even desirable. It is an important part of what makes science scientific, an aspect of the method that Thomas Kuhn calls "the essential tension." (Kuhn, 1977, p. 225.) New ideas come from old ideas, old ideas that have been taken seriously and treated with respect. Furthermore, the process of doing science is structured and disciplined, and this requires some established understanding of nature to specify the structure and give reason for the discipline. Thus there is an unavoidable tension between old ideas and new, between accepting and challenging what you learn in the textbook, between looking things up and thinking for yourself.

Scientific pedagogy, that is, the university education of future scientists, reflects the essential tension in its structure. Kuhn points out, "The single most striking feature of this education is that, to an extent totally unknown in other creative fields, it is conducted entirely through textbooks." (Kuhn 1977, p. 228.) As noted in Chapter 1, other disciplines, like history or sociology, have their students reading primary sources and working with the data early in their education, but the natural-science student is usually kept to the textbooks well into graduate school. There is just so much groundwork to be done, so much background to learn, that it takes these years with the books to develop the expertise. Consequently, the natural sciences are the most dependent on the world on paper in the context of pedagogy. If we were to burn the books and buy stout shoes, the result would be thundering anarchy.

The point here is that some dependence on the authority of established science is both normal and necessary for the advancement of science. It is a large component of what Kuhn calls "normal science" (Kuhn, 1996, p. 10), and it is the structural adhesive of what he calls a paradigm. Authoritative scientists are the spokesmen for the paradigm, the peers doing the peer review, and their authority must be earned. It is earned by their own educational credentials, and these come by the authority of other scientists. It is earned by doing successful and important experiments, sanctioned by the standards of the paradigm. Authority as a scientist rarely follows from actions that threaten or challenge established ideas.

In the case of the tower argument, opposition to the Copernican claim that the Earth moves cited Aristotelian authority on not only the theory about natural motion but even on the outcome of the experiment on the ship. No one did the experiment because it was, apparently, already done and the results were on record, in the books, the world on paper. The books reported that the stone fell some distance behind a moving ship. This was, of course, a repeatable experiment, so it fulfilled that condition of scientific method, but it seems not to have actually been repeated. Again, this is not unusual in science. Journals, more worlds on paper, report experimental results with sufficient detail that the procedure *could* be repeated, but it rarely is. The tower argument is an egregious case of failure to repeat, but arguably not so huge a failure as to be unscientific.

Given the advice to turn to the sensible world, one would expect Galileo to demonstrate the results of the ship experiment by actually doing the experiment. He didn't. Instead he presented another analogy, one that can be done entirely as a thought experiment. This way he avoided the uncertainties that always linger with empirical data, particularly measurements made on the windy, rolling deck of a ship at sea. And, since the result was a matter of reasoning, he showed that it's not just that the stone *does* have a horizontal natural motion, it *must* have a horizontal natural motion.

The thought experiment is easy, perhaps deceptively so. Any analogy or simplifying model is a legitimate demonstration of the facts only if it is appropriately similar to the real thing. Any differences or the details left out must be irrelevant. These points of comparison have to be addressed explicitly, and here again is an important role for the scientist's background knowledge and the authority of existing knowledge. To figure out what happens to the stone on the ship, Galileo said to think about a ball rolling on a flat surface "as smooth as a mirror." Don't worry about air resistance or friction. What happens if the surface is perfectly horizontal and you give the ball an initial push? It will continue to roll without slowing down. But this is just what happens on the moving ship. The horizontal motion of the stone will continue even after the stone is released. The stone will move horizontally with the same speed it had before, that is, the same speed as the ship, and fall at the foot of the mast. And so it is with the tower. Even if the Earth is moving, the dropped stone will fall straight down, as is observed. This follows as long as we can discount the differences between an idealized ball on a mirror and a real stone being dropped. And the legitimacy of that discount must be underwritten by a clear theoretical understanding of the real situation.

With the thought experiment of the ball on the mirror, Galileo introduced what we recognize as the concept of inertia. The object continues to move at its constant speed as long as it experiences no interference, no external forces. But Galileo has not broken from the Aristotelian tradition of natural motion. He has simply added a new natural motion to the stone. It has the natural motion down, and the natural horizontal motion. In fact, the horizontal motion is not straight; it's circular. In other words, the terrestrial object has the celestial natural motion of circular orbit. The ship was moving on the circular surface of the Earth, and that's the path the dropped stone will naturally follow. Galileo has combined the two natural motions in one object. He has linked celestial (circular) properties with terrestrial (straight down), and in this way brought astronomy to physics. You have to wonder, though, if the stone moves naturally both down and around, why don't the planets, which move naturally around, also move naturally down?

Galileo's thought experiment with the ball on a mirror had a much broader consequence than just the tower argument and the movement of the Earth. It led to the Principle of Relativity. Since everything will follow along with a moving reference frame, as the ball continues to move along the mirror and the stone follows the ship, any experiment will have the same outcome in a moving system as in one that is stationary. No experiment can detect the motion of reference frame, as long as that motion is uniform. The tower argument, like any argument that uses evidence from events in the reference system itself, will *in principle* be inconclusive as to whether the system is moving or not. Uniform motion cannot be detected from within.

Galileo's contribution at this point was to show that the tower argument, like the test for stellar parallax, did not unambiguously disprove the Copernican claim that the Earth moves. He went on to provide positive reasons that the Earth does move, first among them his explanation of the tides. Unfortunately, his account of the tides can't be right, as it violates his own Principle of Relativity. Galileo dismissed Kepler's mystical role of the distant Moon as the cause of the tides, and referred instead to the more immediate motion of the Earth itself. More accurately, it is the Earth's *two* motions combined that he said caused the high and low tides. Thus, the oceans' periodic rise and fall is evidence that the Earth moves in just the ways described by Copernicus. This should have been suspect to Galileo from the start, since he used evidence from events in the reference system itself to prove that the system is moving. In-system evidence is *in principle* inconclusive. The principle, of course, is his own Principle of Relativity.

Galileo's argument was this. The Earth has two motions, the orbit around the Sun and the rotation around its own axis. In the not-to-scale Figure 6.1, the Sun is at S and the Earth is the labeled rotating circle. At the point on the Earth marked M (for midnight), the rotational velocity is in the same direction as the orbital velocity, and so these add together to given the resultant velocity of that point. At N (for noon), the rotational velocity is in the opposite direction, and so it subtracts from the orbital velocity. Thus, N is going slower than M. The result is that as one point on the Earth moves around, it must periodically speed up and slow down. The oceans, like water in a basin or barge that moves with uneven speed, will slosh back and forth, up one side and down the other. This creates the tides.

Galileo's explanation of the tides is no proof that the Earth moves, and not only because it is based on the fallacy about uneven speeds in the Earth's rotation. Explaining an observed phenomenon in general doesn't prove that a theory is true. Galileo knew this, and that may be why he offered several reasons to believe

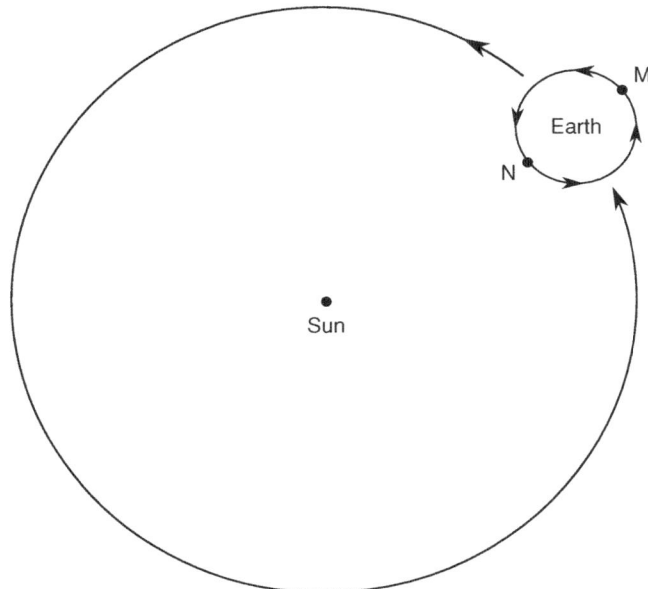

Figure 6.1. Galileo's explanation of the ocean tides, and evidence that the Earth moves. The size of the Earth is greatly exaggerated in this picture. The Earth orbits the Sun and rotates on its axis. The combined motions cause the tides. The point M (for midnight) on the Earth is, according to Galileo, moving faster than the point N (noon). This is because the Earth's two motions add together at M but are in opposite directions at N. The result is a periodic speeding-up and slowing-down of points on the Earth, causing the water in the oceans to slosh back and forth.

the Earth moves. No one of them proves the point with certainty, but several of them together will increase the likelihood. This represents a significant change from the Aristotelian requirement of certainty in science. It is also an important part of, if not the scientific method, then at least the scientific attitude.

By explaining the tides without reference to the distant Moon, Galileo avoided the action at a distance proposed by Kepler. This was conservative science, avoiding the extravagant and unobservable. Galileo allowed a methodological unification between astronomy and physics, but he prevented any physical interaction between celestial and terrestrial objects. The unification that would be universal gravitation was not quite complete.

The telescope helped with the unification, by showing that things in the sky, the Moon and Jupiter, for example, had properties like things on the Earth. Introducing the new tool into the traditional discipline of astronomy was not easy. Indeed, instrumentation in science, and producing images of otherwise unobservable objects and details, should require an account of how the instrument works and why it is accurate. Galileo had little to offer in terms of optical theory. But the telescope worked on land, as one could check by comparing the magnified image to the actual object. From land one could see an approaching ship with the telescope long before it was visible to the naked eye, but when it did arrive the ship looked the same as it had in the telescope. Well, almost the same. The glass used in the lenses of these early telescopes was not entirely clear nor free of bubbles, and so the image formed was not entirely free of distortion.

The bigger challenge to accepting telescopic data came from the celestial/terrestrial distinction. Here was a tool of physics being inappropriately applied to astronomy. There was no way to compare the image of the moons of Jupiter to the actual objects, no way to verify the accuracy without a reliable understanding of how the device worked. And there were real reasons to doubt. A student of Kepler's was anxious to try, but disappointed in the results. "I tested the instrument of Galileo's in a thousand ways, both on things below and on those above. Below it works wonderfully; in the heavens it deceives one, as some fixed stars are seen double." (Quoted in Feyerabend, 1993, p. 88.) The instrument appears to produce astronomical features on its own, turning single stars into doubles. Those mountains on the Moon may be artifacts of distortion as well. A genuinely skeptical scientist should demand either a clear account of how an imaging device works, or a clear demonstration of its accuracy. The astronomical telescope, when first deployed, had neither.

Back on the Earth, and back at work predicting the trajectory of a projectile such as a cannon ball fired in battle, Galileo turned to the careful description of an object in free-fall. From the tower argument he concluded that the stone has two natural motions, one horizontal, that is, circular, and the other vertical. The horizontal motion is easy to keep track of since it remains constant as long as there is nothing pushing or pulling. The vertical motion is more challenging, since the stone clearly speeds up as it falls, and, according to Aristotle and most casual observations, heavier things fall a bit faster than lighter.

This last point could be tested by dropping two objects, one heavier than the other, and seeing if they hit the ground at the same time. But the test would always suffer the uncertainty of imperfect measurements and, much more troublesome, mitigating influences. Air resistance. As with dropping the stone on a moving ship, one would have to interpret the actual results in a way that compensates for uncontrollable, but irrelevant, parameters. If there was no air resistance, the heavy object and light object would hit the ground at exactly the same time. This argument continues the Aristotelian tradition of distinguishing natural and violent motions. The free-fall is natural. Air resistance is violent, an impediment to the natural motion. To learn about the natural vertical motion, we need an experiment that eliminates the violent interference. As he did with the experiment on the ship, Galileo finessed this difficulty by resorting to a thought experiment in which all conditions could be controlled and the one principle could be demonstrated. The one principle turns out to be the Principle of Equivalence.

Galileo's proof was in the form of a *reductio ad absurdum*, or, in the language of mathematicians, an indirect proof. He assumed, for the sake of argument, that Aristotle is right, and heavy things fall faster than light. Consider two objects, a heavy one M and a light one m. By Aristotle's theory, M will fall faster than m. Now consider making a composite object Mm by attaching M and m together. Mm weighs more than M, so, again by Aristotle's theory, Mm falls faster than M. But wait. When you attach m to M, m will retain its inherently slower speed and act as a brake to slow M down. So, again by Aristotle's theory, Mm will fall slower than M. Aristotle's theory, the claim that a heavy object falls faster, leads to a contradiction, that Mm will fall both faster and slower than M. This is impossible, logically impossible, so it can't be that a heavy object falls faster than a slow object.

As with the Principle of Relativity, because Galileo demonstrated that all objects not only *do* fall at the same rate but they *must*, he established a necessary uniformity in nature and not just an empirical generalization. It's not just that nature happens to be this way; it has to be this way. It's a matter of principle. What goes up must come down, and it must come down no faster or slower than anything else.

This is a qualitative law about the natural motion of objects in free-fall, that they all fall at the same rate. Galileo also figured out how to measure what that rate is. He found the quantitative law as well. The challenge was that things fall really fast, making it difficult to time the fall. Galileo devised a way to slow the rate of fall to measurable speeds. This is a remarkable achievement in the development of scientific method, the controlled experiment. The idea is to create an artificial phenomenon that is relevantly similar to what you are interested in in nature. It must be that the differences between the experimental conditions and natural conditions are irrelevant to the phenomenon under investigation. This will require some preliminary theoretical understanding of the phenomenon to know what can be changed and controlled without distorting the interesting results.

Rather than measure the speed of a freely dropped stone, Galileo measured the speed of a ball rolling down a gently sloped incline. This greatly slowed things down, but in a uniform and calculated way. Actually, he measured the position of the ball as a function of the elapsed time from its release at the top. With reference to Figure 6.2, he found the distance s down the slope to be proportional to the square of the elapsed time. In our modern algebraic terms, $s \propto t^2$.

The constant of proportionality between s and t^2 depends on the angle θ of the incline. But the important thing is that it's a *constant* of proportionality, and when $\theta = 90°$, the ball is in free-fall and still the distance it falls is proportional to the square of the time it falls. This general conclusion about vertical natural motion depends on some important details of agreement between the controlled

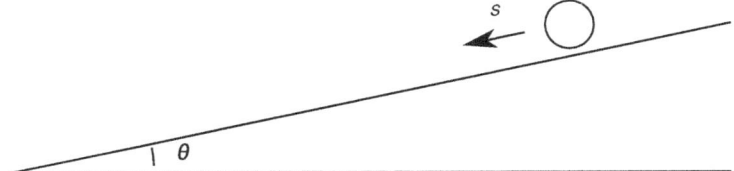

Figure 6.2. The inclined ramp used by Galileo to study the changing speed of an object in free-fall. The distance the ball rolls down the ramp is measured at equal time intervals. The result is that the distance is proportional to the square of the time, $s \propto t^2$. The constant of proportionality depends on the angle θ. When $\theta = 90°$ the object is genuinely in free-fall.

experiment and uncontrolled nature. Air resistance can slightly change the rate of free-fall of objects, but apparently not enough to change the mathematics. Galileo had to argue that we can neglect the effects of air resistance during his actual inclined-plane experiments. Physics students today are used to this, working with a "frictionless surface" during labs and homework. This is, of course, an idealization, and it is exactly what was needed to apply mathematics to physics. The perfect and uncompromising rules of mathematics can describe the messy and complicated events on the Earth if we ignore the mess and complication. This is the genius of experimentation, to decide what can be ignored and measure only what applies to the phenomenon under investigation. That is easy to say, but it takes significant understanding of the situation to know what can legitimately be ignored and what can't. The experiment is an idealization, but it must retain the relevant features of the natural phenomenon. In Galileo's experiment, air resistance can be ignored. So, apparently, can the difference between the constrained motion on the incline and genuinely natural motion in free-fall. The rotational motion of the ball as it rolls down the slope must also be unimportant. And so on. Even the simplest controlled experiment differs in a variety of ways from the natural phenomenon it is meant to demonstrate. The experiment is meaningful only if these are differences that don't make a difference. The facts, in other words, don't speak for themselves; they must be properly interpreted.

Galileo then put together the results of the inclined-plane experiments with the concept of horizontal inertia. He used this to clarify the kinematics of a projectile, an object given an initial push but then left to follow its natural motion, both horizontal and vertical. It will continue its constant horizontal speed, but will fall with increasing speed in the vertical direction in a way that the vertical distance increases as the square of the time. Horizontal distance is proportional to the time; vertical distance is proportional to the time squared. If you plot the position of an object launched horizontally, that is, mark where it is after one second, then two, then three, and so on, the trajectory is the curve shown in Figure 6.3.

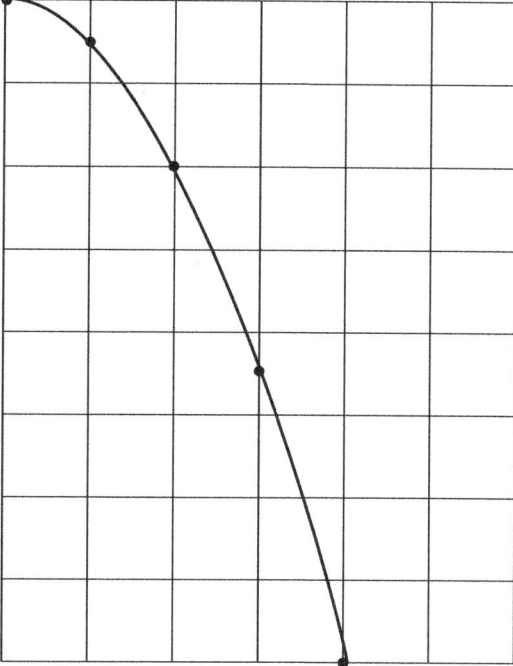

Figure 6.3. Several moments at equal time intervals during the flight of a horizontally thrown projectile. The horizontal speed is the same at each moment, so the horizontal distance between points is always the same. The vertical speed steadily increases, so the vertical distance between points gets progressively longer. Since the vertical distance is proportional to the square of the time, the resulting trajectory is a parabola. This is identical to the path of a projectile in Figure 3.4.

This trajectory is identical to the one in Figure 3.4, the parabolic path calculated from modern, Newtonian gravitation. Galileo did not have the algebraic language to express the formula describing the trajectory, but he got the geometry right. A parabola is a conic section, like an ellipse. Kepler described the motion of planets in terms of one ideal geometric shape, and Galileo described the motion of things on the Earth in terms of another. Kinematics, the description of motion in general, was now in a unified language. With the similarity of kinematics between heaven and Earth, it now made sense to ask the dynamic questions, previously only relevant to terrestrial motion, about celestial motion as well. What goes up must come down, but why? And furthermore, what causes the planets, including the Earth, to orbit the Sun?

Before we see how Newton derived a unified dynamics from this systematic kinematics, it will be worth the wait to first summarize the components of scientific method that have been assembled by Galileo.

Empirical testing of a hypothesis is generally in the form of an if–then statement, and it is never decisive. If the hypothesis is true, then a particular phenomenon will be observed. The hypothesis is not proven by observation of the phenomenon, as the tides don't prove the Earth has two motions. The hypothesis is not disproven if the phenomenon is not observed, as neither Aristotle's tower argument nor Tycho's parallax non-observation disproves the motion of the Earth.

It is worth pointing out, in the interest of understanding both gravity and scientific method, that both Aristotle and Galileo were wrong in what they predicted for the falling stone, dropped from the top of a tower. Without including inertia, Aristotle predicted the stone would fall some distance to the west of the tower. Factoring in inertia, Galileo predicted the stone would fall exactly at the base of the tower. But this was circular inertia. In fact, the natural inertial motion of the stone is straight, tangent to the circular path it was following at the moment of its release from the top of the tower. Knowing this, the prediction is that the stone will hit the ground a bit to the *east* of the tower. This is because the stone at the top of the tower is moving eastward faster than the bottom of the tower. In one revolution of the Earth, the top of the tower goes around a slightly larger circle than the bottom, since the radius of the circle is larger by the amount of the tower's height. The time of revolution is the same, so the top must move faster to cover the greater distance. The dropped stone retains this top-speed throughout its descent – that's inertia – so it passes the east-moving tower. This is an example of the Coriolis effect, the same rotational phenomenon that causes wind to circulate counterclockwise around storms, areas of low pressure, in the northern hemisphere. It's unlikely that Galileo would have been able to detect any eastward advance of the stone. It's too small. If the tower is 100 m high, roughly the height of the San Marco Campanile in Venice, the stone moves about 1.8 cm to the east as it falls.

Science deals with uncertainty. This follows from the logic and conclusions of empirical testing. A false prediction is a challenge, but not an outright disproof. A true prediction is some reason to believe a hypothesis, but not an outright proof. This may be one of Galileo's most valuable contributions to scientific method, the dispensation from Aristotle's requirement of absolute certainty in the results. There will never be just one reason to accept or reject a hypothesis. That decision demands a variety of reasons.

It is not only the logic of empirical testing that shows it to be inconclusive, it is also in the nature of the evidence itself. Scientific observations are not purely objective; they must be interpreted. The interpretation, judging that the evidence is reliable, understanding what it means, and controlling what parameters can be ignored or idealized, must be directed by what is already known about nature. The interpretation is not haphazard or personal. It's *scientific* interpretation, and that means by the standards and guidance of current science. Current theory influences the gathering of evidence. Reciprocally, the gathered evidence influences current theory. Established ideas regulate the introduction and acceptability of new ideas, and news ideas must be allowed to challenge the establishment. This is the essential tension in scientific method. The accomplishment of science will be to build a coherent description of nature that gives authority in equal measure to both evidence and theory, relying on neither as unshakable foundation.

7

Connecting Kinematics and Dynamics

We know where this is going. The astronomy and physics are about to come together in a law of universal gravitation, universal in the sense that a single cause will bring about both terrestrial and celestial effects. From the data and from the theoretical suggestions so far, Isaac Newton (1642–1727) discovered a force that is always exactly along the line between two objects and that decreases as the inverse of the square of the distance between them. These are the key features of the force vector, its direction and its magnitude. The goal of this chapter is to see how Newton got there.

Here is a summary of what Newton had to work with. From Galileo, both the Principle of Equivalence and the so-called time-squared law were key components in the derivation of the law of universal gravitation. Heavy things and light things fall at exactly the same speed, and that speed increases at a constant rate such that the distance fallen is proportional to the square of the time of falling. Both of these describe facts of terrestrial kinematics. From Kepler, the three laws of planetary motion were the most explicitly essential guides to universal gravitation. All three describe facts of celestial kinematics. It was Newton's accomplishment to pull all these together with a unifying force.

Newton himself provided some of the pieces of the puzzle. He, and independently Christiaan Huygens (1629–1695), determined from geometrical considerations that the magnitude of the acceleration of an object moving in a circle, the centripetal acceleration, depends on the square of the speed and inversely on the radius of the circle. This is a very general result that applies to circular motion on the Earth or in the sky. In this sense it is universal kinematics.

$$a = v^2/r \tag{7.1}$$

Newton also made the essential link between kinematics and dynamics with the second law of motion.

$$F = ma \tag{7.2}$$

In both of these equations we are describing only the magnitudes of the force and acceleration. So, even though they are vectors, we write the math without putting the variables in bold.

Equally important was Newton's third law of motion, that forces always come in pairs. If one object, the Sun, for example, exerts a force on another, a planet, then the planet exerts a force of equal magnitude on the Sun and exactly along the line between the two objects. This will bring some subtle but important changes to Kepler's model of planetary motion.

These were the ingredients. The challenge was to put them together into a theoretically coherent law that accurately matched the empirical data. As a general note on the logic of reasoning in this kind of situation, it is relatively easy to figure out how things will move if you already know the details of the cause of the motion and the initial conditions. This is the inference from dynamics to kinematics, from force to resulting acceleration. On the presumption that nature is deterministic rather than random, a cause determines the effect. But even in a deterministic world, a particular effect could have more than one possible cause. There may be more than one way to bring it about. This means that any inference from kinematics to dynamics is difficult. It is **underdetermined** in the sense that all the descriptive details on how things are moving do not have enough information on exactly what is causing the motion. Kinematics underdetermines dynamics, but this is exactly what Newton needs to do, "from the phenomena of motions to investigate the forces of nature." And he is not alone. Again from the Preface to the *Principia*, "All the difficulty of [natural] philosophy seems to consist in this." Figuring out the cause from the nature of the effect is a fundamental challenge, and accomplishment, of science. When it is explicitly presented in these terms of causation and inference it is called an **inverse problem**. This is to distinguish it from a direct problem, predicting the effect of a particular cause.

Newton's inverse problem was solved by first ignoring some of the details in the data. Just as Galileo was able to apply mathematics to experiments by judiciously discounting certain small imperfections in the phenomena, Newton had to idealize on the astronomical facts. The initial derivation of the law of universal gravitation assumed that the orbits of planets are perfect circles. It ignored, in other words, Kepler's first law. Remarkably, by the end of Newton's development of the theory of gravitation, the assumption of circularity disappeared, and the mathematics resulted in elliptical orbits. In this way the system of ideas that tied all the kinematic and dynamic pieces together was self-correcting.

Newton also started out neglecting any movement of the Sun. This too changed as the theory was fully developed, but initially the Sun's overwhelming mass allowed Newton to consider it unmoved by any of the planets, individually or collectively.

And finally, Newton regarded the Sun and each of the planets to be an amount of mass concentrated at a single point. He neglected the physical extension of the astronomical objects. He returned to this idealization after the fact, and, using the calculus, was able to explicitly prove that the gravitational effect of a spherical body is identical to that of a point particle. So, this idealized detail in the original derivation was not *self*-correcting, but it was harmless.

The next few pages include a few mathematical steps in Newton's derivation of universal gravitation. You can skip the math and still get the basic logic by skimming ahead to Equation (7.9) and the paragraph that starts, "To summarize the logic..."

With these initial simplifications, Newton was first able to link Kepler's third law to the inverse-square decrease in the force of attraction between two astronomical bodies. If the ratio of the orbital radius cubed to the period squared is the same for every planet, it follows that the magnitude of the force between the Sun and a planet decreases by the inverse-square law. In a succinct logical form, if $r^3/P^2 = a$ constant, then $F \propto 1/r^2$.

Here is a sketch of the proof.

Considering the planetary orbit to be circular, the planet's acceleration is, by Equation (7.1), v^2/r. By Newton's second law, Equation (7.2), this requires a force that is the mass times the acceleration.

$$F_{\text{gravity}} = mv^2/r \qquad (7.3)$$

The speed v is the distance the planet travels per time, namely the circumference of its orbit divided by the time it takes for one complete orbit, the period P. The circumference is $2\pi r$, so the required force is as follows.

$$F_{\text{gravity}} = m(2\pi r/P)^2/r \qquad (7.4)$$

Multiply the right-hand side by r/r, that is, by 1, and regroup.

$$F_{\text{gravity}} = (1/r^2) \times [(4\pi^2 m) \times (r^3/P^2)] \qquad (7.5)$$

Everything in the square bracket is a constant, including the (r^3/P^2), since that is the constant specified by Kepler's third law. The result is that the force on the planet is some constant times $1/r^2$. In other words, we have the following.

$$F_{\text{gravity}} \propto 1/r^2 \qquad (7.6)$$

The next step was to apply this idea to the orbit of the Moon, and compare the force holding the Moon in orbit to the force that accelerates a dropped stone or a falling apple. For this comparison, assume all of the mass of the Earth is located at a single point at the center. The distance to the Moon, the radius of its (approximately) circular orbit, is measured from the center of the Earth to the center of the Moon.

Using our current value and units, this distance is $R_M = 3.84 \times 10^8$m. We'll use the capital R to denote a specific value of distance between two specific things, in this case the Earth and the Moon. Lower case r will be used for general, variable distance between unspecified objects. The dropped stone on the surface of the Earth is at a distance from the point source of gravity that is just equal to the radius of the Earth itself, R_E. That is, the stone is just like the Moon, left to its natural motion, but at a distance $R_E = 6.38 \times 10^8$m. The ratio of the two distances, R_E/R_M is 1/60.

Since it moves in a circular orbit, the Moon accelerates toward the center at a rate of v^2/R_M. The period of its orbit is roughly 28 days, or more precisely, 2.3×10^6 s. When you do the math, the Moon's acceleration is $a_M = 2.7 \times 10^{-3}$ m/s^2. The stone, or any other object dropped on the surface of the Earth, accelerates toward the center at a rate of 9.8 m/s^2, the acceleration of gravity g. The ratio of acceleration on the Earth to acceleration of the Moon is $g/a_M = 3600$. Note, as Newton did, that this is the square of the inverse of the ratio of distances. And, since there is already reason to believe that the force holding the Moon in orbit depends on the square of the inverse of the distance, and that acceleration is proportional to the force, it follows that it is the same force holding the Moon in orbit as is the cause of the stone's falling. Thus gravity is universal, and astronomy and physics are united.

The acceleration of gravity, the rate of falling on the Earth, does not depend on the mass of the falling object. The only way this is possible is if the mass appears on both sides of the equation $F = ma$. It has to be on the left, in the formula for F, in order to cancel with its appearance on the right. So we now have two components of the force of gravity, mass m and $1/r^2$.

$$F_{\text{gravity}} \propto m/r^2 \qquad (7.7)$$

The force holding the Moon in orbit, the force of gravity, is proportional to the mass of the moon M_M (again using the capitalized notation to signify a specific value of a specific object) and the inverse of the distance squared $1/R_M^2$. Now use Newton's third law to point out that there must also be an equal and opposite force on the Earth caused by the Moon. By this symmetry we could have done the derivation from the perspective of the Moon, asking about the acceleration of the Earth. That puts the mass of the Earth, the accelerating object, on the right-hand side of the $F = ma$ equation. And again, to make the acceleration independent of the mass, that factor has to be on the left-hand side of the equation as well, that is, in the formula for the force. Both masses, those of the Earth and the Moon, have to be in the formula for the force of gravity. This is required by the Principle of Equivalence. In general, for any two masses m_1 and m_2 separated by a distance r, we have the following.

$$F_{\text{gravity}} \propto m_1 m_2/r^2 \qquad (7.8)$$

All that's left is to determine the constant of proportionality, the so-called gravitational constant G. There is no principle or law that can be used to derive this constant. It is an empirical matter. It turns out to be $G = 6.7 \times 10^{-11} \mathrm{m}^3/\mathrm{kg\ s}^2$. All together, Newton's law of universal gravitation is in the following formula.

$$F_{\text{gravity}} = G m_1 m_2 / r^2 \tag{7.9}$$

There were more steps in Newton's derivation, but they can be summarized without the mathematical detail. He demonstrated that a central force, that is, a force directed exactly along the line between planet and Sun, results in Kepler's second law, that the planet sweeps out equal areas in equal times. (For this proof, see Cushing 1998, pp. 115–117.) This works for any central force, not just one with a $1/r^2$ dependence. When you add the $1/r^2$, the resulting orbit turns out to be an ellipse. (Again, Cushing 1998 gives a clear account of this, on pp. 119–122.) A central, inverse-square force, in other words, entails Kepler's first law. And though Newton started with the approximation of circular orbits, his analysis ended up with ellipses.

To summarize the logic of Newton's solution of the inverse problem, his derivation of the law of universal gravitation, start with Kepler's third law and Newton's own $F = ma$. From these the inverse-square relation, the $1/r^2$, follows. And, when combined with the Principle of Equivalence, that inertial mass is the same thing as gravitational-source mass, and Newton's third law of motion, it follows that the force on the Moon, and other orbiting objects, is the same as the force on a falling apple or stone, and it depends on the masses of the two bodies, Earth and Moon, or Earth and apple. The force of gravity is universal. Finally, combine the inverse-square relation with the central direction of the force, and both Kepler's second and first laws follow. The orbits of planets are elliptical.

The orbits of the planets are elliptical, but there is a subtle difference between Newton's model and Kepler's. It comes from Newton's third law, the one about equal and opposite forces. The Sun exerts a force on a planet to hold it in orbit and cause it to accelerate, so the planet must exert exactly the same force on the Sun. The Sun must be accelerating and in orbit. The rate of the Sun's acceleration is much less than that of the planet, since the Sun is more massive. We could use the orbit equation (Equation (3.7)) to show that this means the radius of the Sun's orbit is much smaller than that of the planet. The Sun and planet orbit the same point, the center of mass between the two. This point is at the focus of the ellipse. In other words, the planetary orbit is, as Kepler reported, elliptical, but the Sun is not at the hearth. The planet orbits an empty point in space, a void point, as shown in Figure 7.1.

There is an important difference between the void point in the Newtonian model and void points in previous models. The location of the void point for Newton is determined by the locations and properties, specifically masses, of real objects.

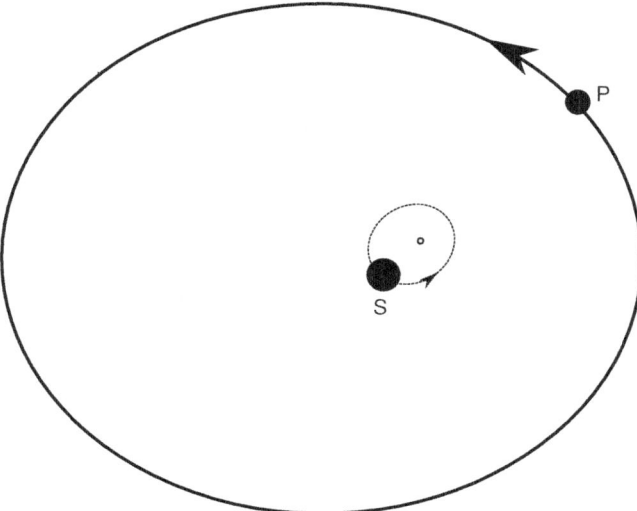

Figure 7.1. The Newtonian model of planetary motion. This is Kepler's model, revised as necessary following Newton's laws of motion. Both the planet P and Sun S are in motion. They follow elliptical orbits around a common focus. The planet's orbit is much larger than the Sun's, but the orbital period is the same. The focus of the orbits is at the center of mass between S and P. It's a void point, but its location is determined by the masses of the real physical objects S and P.

The void point is at the center of mass of the objects involved. This is unlike the Aristotelian situation, in which the possible void point is simply at the center of the universe, independent of what the universe contains. And it is unlike the Ptolemaic void points that are strictly kinematic, with no dynamic role or consequences, and whose position and motion are freely adjustable to match the appearances of the planets, that is, to save the phenomena. In the Newtonian model, it's not the point in space that is having the dynamic effect; it's the mass of Sun and planet balanced at the unoccupied point in space. So, it's a void point, and a dynamic void point, but it is allowable by Machian principles that require real things to be the physical causes of motion.

Newton solved the first part of the basic problem of natural philosophy, "from the phenomena of motions to investigate the forces of nature . . . " He, and his successors, went on to great success with the second part as well, "from these forces to demonstrate the other phenomena." The law of universal gravitation proved to be useful in explaining a variety of happenings in nature, and in getting a variety of things done. Its applications were genuinely universal.

The Newtonian theory explained the ocean tides. The mathematics are a bit daunting, but the basic mechanism is not. We can understand what's going on by using the modern concept of a field. This was not available to Newton, but with a

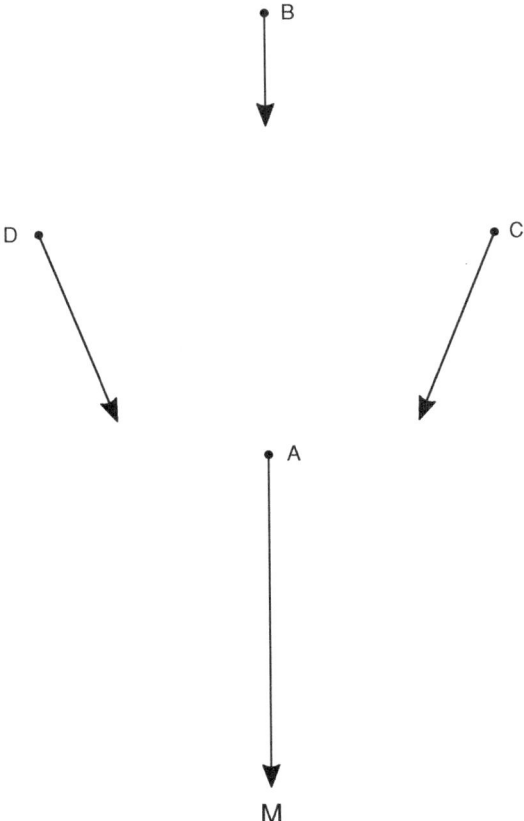

Figure 7.2. The gravitational field produced by a mass source M. The field is radial, meaning that all of the field vectors point directly toward M. The strength of the field decreases at greater distance from M. These two features make this an inhomogeneous field; the magnitude and orientation of the vectors are variables of position.

few basic field vectors, directly from his formula for the force of gravity, the tidal phenomenon is apparent. Figure 7.2 shows a few of the vectors of gravitational force on a test mass at various places near the gravitational source M. The important features are the decrease in length (magnitude) of the force at greater distance from M, and the radial orientation of the vectors; they all point exactly toward M. These properties follow from the facts that the gravitational force decreases as $1/r^2$ and is a central force. In the language of field theory, this is an inhomogeneous field, since the direction and magnitude of the vectors are different at different points in space.

Consider four small test particles, each of mass m, at each of the points A, B, C, and D in Figure 7.2. A falls faster than B, since it is closer to the source M, so

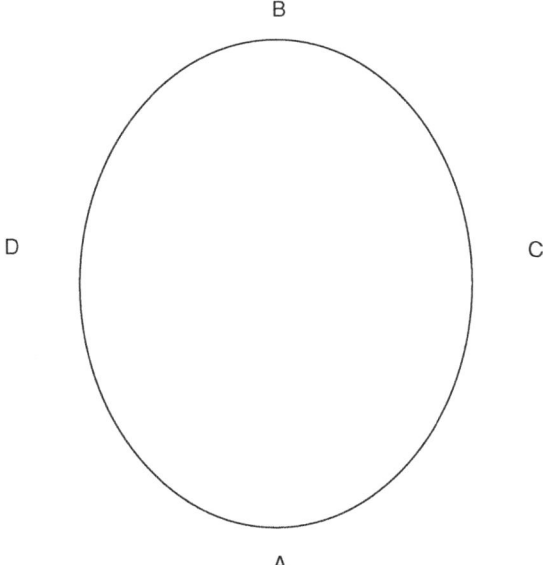

Figure 7.3. The distorting effect on a fluid in the inhomogeneous field of mass M. This is a tidal effect, and the oceans on the Earth will bulge out at points A and B, in at points C and D, in the gravitational field of the Moon. There will be high tides at A and B, low tides at C and D.

their tendency is to separate, to get further apart. C and D fall at the same rate, but they each have a component of force toward the other. Their tendency is to move closer together, squeezed by the converging, inhomogeneous field. In free-fall, this quartet would bulge out along the AB fall-line, and squeeze in along the CD line. If it were a water balloon in free-fall in this inhomogeneous field, it would in fact elongate, as in Figure 7.3.

The Earth is in free-fall toward the Moon. The Earth has a gravitational force on the Moon, and by Newton's third law the Moon has the equal and opposite force on the Earth. Like the Sun and planets, the Earth and Moon each orbits the center of mass between them. So, the Earth is in orbit, and that is simply free-fall – the only force acting on the Earth in this system is the gravity of the Moon – with sufficient tangential orbital velocity to avoid hitting the Moon. Fluids, like water, on the Earth will react just as the fluids in the freely falling water balloon. They will bulge, rise, at points closest to and furthest from the gravitational source M (now the Moon), and they will sink inward, lower, at points in between, C and D. High tides at A and B, low tides at C and D. With the Earth rotating as well as orbiting in the Earth–Moon system, those points are not stationary on the ground. Just as the position of the Moon changes in the sky, so do the positions of high and low tides.

Newton's law of universal gravitation seems to be on to something fundamental in nature, since it can, with a single simple formula, account for such diverse phenomena as the orbits of planets, falling apples and stones, the ocean tides, and our own feeling of weight. It was a remarkable accomplishment. But it came at considerable metaphysical cost, requiring a dynamic connection between two things with no contact and even no intermediate physical stuff between them. This is the action at a distance that Galileo and then Mach found mystical and (hence) unscientific. But it works, and that may be enough to make it acceptably scientific. It works especially well when it is conceptually upgraded to being a thing itself, a field.

Another application of the law of universal gravitation highlights just how effective and useful it is to think in terms of fields. As a consequence of the $1/r^2$ in the force of gravity, it is possible to escape the gravitational bonds of the source such as the Sun or the Earth. Since the force decreases quickly with distance from the source – it's the inverse of the distance *squared* – it's possible to start out going fast enough that you never slow down to zero and fall. What goes up, doesn't come down, as long as it starts going up fast enough. The particular speed required is the **escape velocity**. It depends on the mass of the object you are leaving, the Sun or a planet, say, and where you start. The closer you are to the source, and the more massive the source, the more speed it will take to get away.

It's not difficult to calculate exactly what the escape velocity is, as a function of the mass and distance of the source. Then we can put in the numbers, for example, the mass and radius of the Earth, to see how fast you would have to go to escape the Earth's gravitational pull. It's particularly easy using fields, and the concept of energy.

The kinetic energy K of an object is defined as $K = \frac{1}{2}mv^2$. It is the energy of movement, and something that is not moving, $v = 0$, has no kinetic energy. Kinetic energy is a scalar, and this is what makes it easy to work with. A related property is work W. It's related in that, as defined in physics, work also requires that something moves. It is also a scalar. The work done by a force F is defined as $W = Fx$, where x is the distance the object moves under the influence of the force. This is a simplified definition; it works only if the force is constant and the motion is exactly along the direction of the force. It can be generalized to variable force and any orientation of motion, but this basic case will serve.

If you hold a heavy stone at arm's length, no work is done. You are indeed applying a force to oppose the gravitational force on the stone; you have to push up with a force equal to the stone's weight, $F = mg$. But since there is no change of position, no x, there is no work done. If you *lift* the stone through a vertical distance x, then you do work by an amount Fx, that is, mgx. If you apply a force horizontally

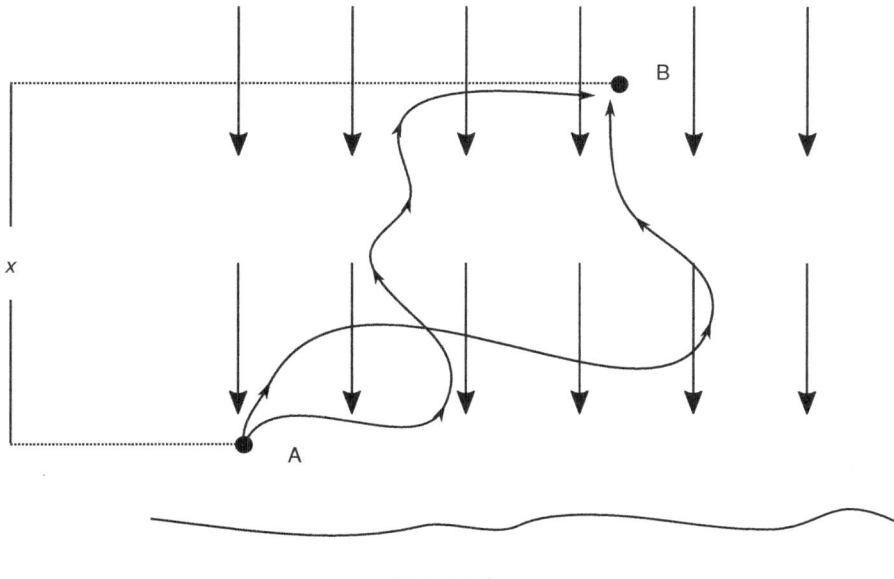

Figure 7.4. The potential energy difference between points at different heights in a homogeneous gravitational field. The potential energy at B is greater than the potential energy at A by an amount that depends only on the vertical distance x (and the mass of an object at point B and A). It is independent of the path taken from A to B.

to an object, that will cause it to accelerate. This work changes the velocity of the object, and consequently changes its kinetic energy.

Focusing on this last feature of work, the defining formula can be rewritten to make it explicit that the work done on an object is equal to the change in its kinetic energy ΔK. That is, $W = \Delta K$. But what happened in the case of lifting the stone? You did work to move the stone some vertical distance x, but its speed, and hence its kinetic energy, did not change during the lifting. To cover this, we expand the idea of energy. There is kinetic energy, the energy of motion, and there is **potential energy** V, a kind of stored energy that can be released to do work. In the lifting case, the work increases the potential energy by positioning the stone higher in the gravitational field. If released from that height, the stone would pick up speed, and the work done pays off in the form of kinetic energy. Potential energy is energy in terms of position in the field. Higher up in the gravitational field means greater potential energy. And it doesn't matter how the object got higher up. The gravitational field near the surface of the Earth is approximately homogeneous, as shown in Figure 7.4. The change in potential energy ΔV between points A and B depends only on the vertical distance x between the two points: $\Delta V = mgx$. The total work done to get from A to B is path-independent. This is one of the simplifying virtues of working with scalars. The result depends only on the end-points.

The most general relation between work and energy is that the work done by a force is the change in *total* energy E, kinetic plus potential. If there is no external force applied and hence no work done, then the total energy is constant. This is the law of conservation of energy.

Apply the law of conservation of energy to an example of tossing a stone straight up. What goes up, must come down. How fast will it be going when it comes down? If it starts with speed v, say 10 m/s, then it will return with exactly the same speed, 10 m/s. As it goes up, the kinetic energy decreases while the potential energy increases. The total energy stays the same. At the top of its flight, the stone's vertical speed is zero in the instant it must stop and then head down. At the top, all the energy is potential energy. As it goes down, the potential energy decreases while the kinetic energy increases. When it reaches the same vertical position from which it was tossed at 10 m/s, all the potential energy will be returned as kinetic. In other words, the kinetic energy will be the same returning as when it was launched, so the speed will be the same. What goes up at 10 m/s will come down at 10 m/s (at the altitude of the toss). And since the calculation is done in terms of scalars, work and energy, this result does not depend on the direction of the throw. Throw the stone straight up at 10 m/s, or at a 45° angle at 10 m/s, and when it returns to the ground it will be going 10 m/s.

Gravity is a conservative force. All the energy put in to move around in the field gets returned when you get back to where you started. No energy is lost, dissipated, as it would be to the force of friction. Friction is a non-conservative force. There is no potential energy associated with the force of friction, and there is no field in which to determine the energy of position. But with gravity and other conservative forces, there is a determinate property of position and path through the field, and so there is potential energy.

The analysis of potential energy has so far been assuming the gravitational field to be homogeneous, as shown in Figure 7.4. This is approximately true for the field near the surface of the Earth or the Sun, but we know that in fact the field is not homogeneous. The force gets weaker with distance as $1/r^2$, and the field lines converge toward the center of the source. The variable magnitude of force means that calculating the work to move an object from one altitude to another is not easy. It requires calculus. The procedure is to figure the work for a tiny distance, over which the force is essentially constant, and then add up all these increments of work from A to B. The general procedure, which we won't do, shows that if the force varies as $1/r^2$, then the potential energy varies as $-1/r$. The minus sign makes sense, since the potential energy *increases* as r increases. Further from the ground

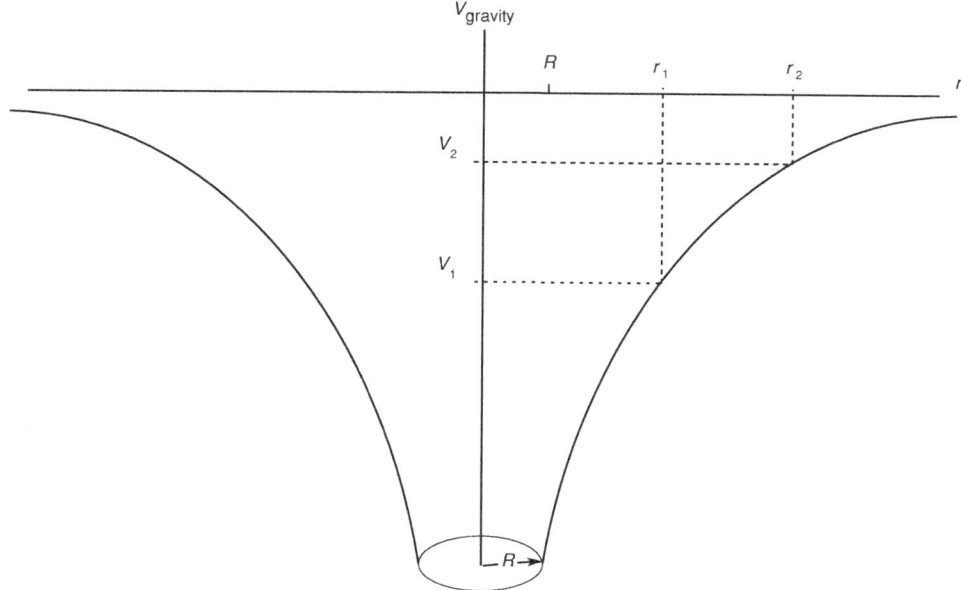

Figure 7.5. A graph of gravitational potential energy V_{gravity} as a function of distance r from the source. The source, a star or planet, has physical radius R, so the graph and the formula $V = -GMm/r$ are only for values of r greater than R. At infinite distance, $r \to \infty$, the potential energy is zero. The graph can be used to determine how much kinetic energy, and hence how much speed, is needed to get from one distance r_1 to another r_2. If $r_2 \to \infty$, the calculated speed is exactly the escape velocity at point r_1.

means more energy stored. Without the minus sign v would be decreasing as $1/r$. The potential energy of a mass m at a distance r from a planet or the Sun with mass M is as follows.

$$V_{\text{gravity}} = -GMm/r \qquad (7.10)$$

A coordinate-system choice has been made in this formula, with the convention that $V = 0$ an infinite distance away from the source. This way, the source is understood to create a potential well, as shown in Figure 7.5, where the potential energy V is plotted as a function of distance r from the source.

If this is the potential well specifically for the Earth, then objects on the ground are at distance R_{E}, the radius of the Earth. An object dropped from an altitude r_1 will have potential energy V_1, as shown on Figure 7.5. Since it starts with zero velocity (it's dropped, not thrown) it starts with zero kinetic energy. Its total energy E is therefore all potential, that is, V_1. When it reaches the surface of the Earth, distance R_{E}, it will have acquired kinetic energy exactly equal to the change in potential energy between the position r_1 and R_{E}. Call this value of kinetic energy

K_1. And knowing the kinetic energy, and the mass of the object, we could calculate the object's speed. K_1 is also the kinetic energy, and associated speed, needed to launch the object from the surface of the Earth to reach an altitude r_1. As it rises above the Earth, it loses speed, and hence kinetic energy, and gains potential energy. The point where the kinetic energy goes to zero, and where the vertical flight stops, is where all the energy has become potential, that is, r_1. To launch a rocket to an altitude r_1, the rocket needs initial kinetic energy such that, as it loses speed, $v = 0$ when $r = r_1$.

Greater altitude requires more energy, greater launch-speed. To get to r_2 from the surface of the Earth requires greater kinetic energy. The rocket has to climb higher up the well. But if had *started* from r_1, to get to r_2 would require only the difference between V_1 and V_2.

The potential-energy well and the law of conservation of energy can be used to derive the exact formula for escape velocity. Think of it as escape kinetic energy. How much kinetic energy is required to escape the well? To just exactly make the escape, the launched object will slow to zero velocity at infinite distance, that is, $v = 0$ when $r \to \infty$. At that point, $K = 0$. By the choice of coordinate system in Figure 7.5, $V = 0$, also at infinite distance. The total energy E is the sum of K and V, so at infinite distance, $E = K + V = 0 + 0 = 0$. Actually, since total energy is a constant, the total energy is zero at *all* points and *all* times in the flight of an object that has exactly what it takes to escape. So, $K + V = 0$, and $K = -V$. This is the energy requirement for escape. Putting in the specifics for K and V, $\frac{1}{2}mv^2 = GMm/r$, and solving for v, we find the escape velocity.

$$v_{escape} = (2GM/r)^{1/2} \tag{7.11}$$

The escape velocity depends on where you start, r, and on the mass of the gravitational source, M. It does not depend on the mass m of the escaping object, the rocket or the stone; m dropped out of the equation, as it usually does with gravity when the equation is kinematics. Gravitation is universal both in the sense of affecting everything everywhere, but also in the sense of being egalitarian in its effect. Whether it's a rocket or a stone, the necessary escape velocity from the Earth is the same. And when you put in the numbers for the mass and radius of the Earth, the radius because we are launching from the Earth's surface, the escape velocity from the surface of the Earth is 1.1×10^4 m/s, that is, 11 km/s.

The escape velocity from any mass source M depends on where you are. The closer to M, the stronger the gravity and so the faster you have to go to escape. This has an interesting consequence for very dense bodies, that is, bodies with a lot of mass M compacted inside a small radius R. It means you can get very close to the mass, so the escape velocity must be very fast, perhaps even faster than the speed of light. You can see where this is going, or maybe you can't. The idea of a black hole, an object so dense that light cannot escape its gravitational field, was suggested not long after Newton derived the $1/r^2$ formula for the force of gravity. It's a pretty simple idea. If the Earth were to shrink in size but not in mass, and you stayed at the same distance r from the center, the escape velocity would remain at 11 km/s. That's because M and r don't change in the v_{escape} formula. But if you rode in on the surface of the shrinking Earth, r would decrease so v_{escape} would increase. If the shrinking continues you will get to a radius so small that v_{escape} equals the speed of light $c = 3 \times 10^8$ m/s. The calculation is easy and, for the mass of the Earth, the radius at which the escape velocity is equal to the speed of light is about 10 mm. All of the mass of the Earth compressed to a sphere that small would create a black hole. This early understanding of black holes is a bit simplistic, in that it ignores some important properties of light, but it gets the basic idea across that gravity can affect even light.

There is more you can do with Newton's law of universal gravitation, the basic $F_{gravity} = Gm_1m_2/r^2$ formula. The orbit equation (Equation (3.7)) that prescribes the orbital speed and radius for any planet, moon, or man-launched satellite is easily derivable. This, and the correlation between gravitational potential energy and kinetic energy – speed again – are the basis of contemporary space travel. That came later, centuries after Newton but, even at Newton's time, the applications of the theory were many and varied. It may be fair to summarize the accomplishment in terms of coherence and simplicity. What had appeared to be very different kinds of phenomena governed by different laws were brought together in a systematic connection between Earth and sky, cause and effect, force and motion. The pieces now fit together in the efficient package of Newton's second law and the law of universal gravitation. These are some of the conceptual virtues, simplicity and coherence, discussed in Chapter 4. Add to these the empirical virtues of explaining so many phenomena and being so useful in getting things done (it *is* rocket science!), and the theory seems well justified by scientific standards. All that remained was explicitly testing the theory in the pattern of precise prediction and follow-up observation. A scientific theory must be both testable and tested. Newton's theory was both, and two of the most informative tests, informative about gravity and about scientific method, will be described in the next chapter.

8

Testing the Newtonian Theory

Open any physics textbook and you will learn that gravity is a force of attraction between two objects that depends on their individual masses and the distance between them. The description will be clear (if it's a good textbook) and precise. It will be in terms of the gravitational field and field lines. The math will be pretty simple, directed through the Newtonian formula, $F_{\text{gravity}} = GMm/r^2$. The tone will be confident, as is the style of textbooks. We know what gravity is, and what it isn't. There will be no talk of an object's proper place in the universe or its goal of getting there. There will be no references to crystalline spheres in the astronomy section, since we now know it's simply this force of gravity that holds a planet in orbit.

There will be no worries in the textbooks about, as Ernst Mach put it, the "uncommon unintelligibility" (Mach, 1911, p. 56) of an invisible field that can exert a force instantaneously, at great distance, between any two objects. Whether or not it has become intelligible, this idea has become common knowledge. The Newtonian theory is the truth about gravity taught to physics students throughout their undergraduate education. It's not until graduate school that the more modern theory, the general theory of relativity, is introduced as the proper replacement for Newton. And even then, a class in general relativity is usually optional, not a requirement for the graduate degree. You can get a degree in physics, surely a bachelor's degree and maybe even a doctorate, with a Newtonian understanding of gravity.

Often the textbooks will hint that there are troubles for the Newtonian theory, and that, strictly speaking, it's not the last word on gravity. But it's surely good enough. It's good enough for most work in physics. It's good enough for all work by engineers. And it's good enough for rocket science.

Good enough means that if you use the Newtonian theory of gravity to figure out how to launch and aim a rocket to get to Mars, the rocket will get to Mars. You can even land it safely on Mars, again by using the Newtonian theory to design and deploy the steering and braking mechanisms. This sort of successful application makes it seem as if the theory is true. Even if it is not spot-on true, it is at least nearly true. It's true enough.

A word of caution is in order. It is certainly possible for a scientific theory to facilitate accurate applications even when the theory is dramatically false. Consider the caloric theory of heat. If you had opened a general physics textbook, or a more specific thermodynamics textbook, in the eighteenth and early nineteenth centuries, you would have learned that heat is an invisible substance, a fluid that flows from hot objects to cold. This accurately predicted that most things will expand as they get hotter, to make room for the extra stuff, the caloric fluid. It was also the theoretical basics for the industrial revolution, the invention and development of steam engines and other practical ways of using heat to get work done. But the theory was, and is, false. Really false. It's based on a fundamental fiction. There is no such thing as a caloric fluid.

We worked with the difference between a theory being true and its being useful in Chapter 4. That was about the Ptolemaic model of planetary orbits, the one with epicycles and void points. The suggestion was that practical applications and success with the evidence are all we should really care about in a scientific theory. The theory being true or not is irrelevant. This attitude toward theories is called instrumentalism, a sort of shut up and calculate response to the question about truth.

Instrumentalism may get you through the work day as an engineer or a scientist, but it conflicts with some inherent human curiosity, the one that wants to know what's really going on in nature. Besides, the textbooks don't say it's *as if* there is a force field; they say *there is* a force field. Instrumentalism is dissatisfying because it denies the natural inference that a theory works *because* it's true. That is, the success with evidence and applications indicates that the theory is basically right. How could it work if it were false? Germ theory is fantastically successful in keeping us healthy, and that's because there are, in fact, germs.

So, what, if anything, is the connection between the utility and the truth of a theory? On the one hand, theories can work really well even if they are false. On the other hand, it seems like a theory couldn't work as well as Newtonian gravity without being true, or nearly true. To sort this out, we need to look carefully at the reasons one might think the theory is in fact true. There are two kinds of things to look at, empirical testing (obviously), and the internal, conceptual workings of the theory itself. This second aspect of evaluating a theory asks about things like the simplicity and theoretical coherence of the ideas. Maybe the indication of truth is in the combination of these things. A true theory is one that works with the evidence and is as simple and coherent as possible. It has to both make sense of the observations and make sense itself.

Does the Newtonian theory of gravity make sense? Is it intrinsically plausible in how it describes the mechanism of gravity? Despite our comfort with the idea of a gravitational field, the initial reception of Newtonian gravity faced considerable conceptual challenge. The problem was the action at a distance. The force between two objects requires no contact or substantive medium. And it happens instantaneously. We live with this by letting the field do the work. This may be just hiding one mystery behind another. The gravitational field is a dynamic player in the gravitational interaction. There is nothing else there between the two. The field is the thing itself that gets the job done.

This action without contact used to be regarded as not only mysterious but mystical. When Kepler suggested that the Moon was the cause of the ocean tides on the Earth, Galileo dismissed this no-contact interaction as occult and argued for his own, local cause. But then Newton allowed the distant force suggested by Kepler, and made it universal gravitation. There was still some objection at the time. For example,

It is inconceivable, that inanimate brute matter should, without the mediation of something else, which is not material, operate upon, and affect other matter without mutual contact; as it must do, if gravitation, in the sense of Epicurus, be essential and inherent in it. And this is one reason, why I desire you would not ascribe innate gravity to me. That gravity should be innate, inherent, and essential to matter, so that one body may act upon another, at a distance through a vacuum, without the mediation of anything else, by and through which their action and force may be conveyed from one to another, is to me so great an absurdity, that I believe no man who has in philosophical matters a competent faculty of thinking, can ever fall into it.

(Bentley, 1838, p. 211.)

This is, remarkably, from a letter that Newton himself wrote to a colleague, Richard Bentley. The mechanism of universal gravitation is an "absurdity." But it works. And we have come to overlook the absurdity. That's how Mach saw it in 1911.

The Newtonian theory of gravitation, on its appearance, disturbed almost all investigators of nature because it was founded on an uncommon unintelligibility. People tried to reduce gravitation to pressure and impact. At the present day gravitation no longer disturbs anybody; it has become *common* unintelligibility.

(Mach, 1911, p. 56.)

Mach's point is that the theory still doesn't make sense, but we have gotten used to it. Complacency is no indication of truth. If we are to regard the theory as true, it will have to be *despite* its not making sense. This will require some pretty robust evidence. The theory will have to pass some strict tests.

In the case of a theory of gravity, and specifically the Newtonian theory of gravity, the testing will have to be indirect. As we realized in Chapter 1, neither the force of gravity nor the field is something we can directly observe. So testing the theory will have to be done by observing something else, something that is logically related to the theoretical ideas about forces and fields. We should further clarify this logical relation, something we started in Chapter 6, before considering the evidence, the observations.

It is often said that a scientific theory is tested by its observable predictions. The evidence is in seeing whether the predictions turn out to be true. There is also favorable evidence if a theory can explain what has already been observed. The logic is the same. In both cases, prediction and explanation, it's an if–then relation between theory and evidence. It's always: If the theory is true, then a specific observation is expected. To illustrate with a simple medical example, if malaria is a mosquito-borne parasite, then we predict that preventing mosquito bites will prevent the disease. And if malaria is a mosquito-borne parasite, then that explains why the disease is most common in warm, wet climates. Explanation or prediction, theory and evidence always meet in an if–then relation. Theory is the if; evidence is the then.

Consider the phenomena that the Newtonian theory of gravity explains. One of Newton's great mathematical achievements was to prove that if the gravitational attraction between two bodies depends on the inverse-square of the distance between them, the $1/r^2$ in the formula, then Kepler's three laws of planetary motion followed. The theory explains Kepler's laws, and the laws themselves are summaries of observations, the regularities in the orbits. The Newtonian theory also explains the ocean tides, again by mathematical derivation of the form, if the theory is true, then the tides as we observe them follow. And the theory explains what is observed in the phenomenon of free-fall, specifically the time-square law of Galileo and the fact that heavy and light objects fall at the same rate.

In all cases of explanation, the argument has two premises.

(1) If the theory is true, then the phenomenon will be observed.
(2) The phenomenon has been observed.

Here's where we have to be careful, because to draw the conclusion that

(3) the theory is true,

would be to commit a logical fallacy.

Logicians call this the fallacy of **affirming the consequent**. The mistake is obvious when you apply the same form of reasoning to diagnosing disease. Think of a case in which the patient has a fever. That's all we know at this point, the fever. A doctor, a bad doctor, may reason that malaria would explain the fever, so the truth of the matter is the patient has malaria. It's true that malaria explains the fever. If the patient has malaria, then they will have a fever. And it's true the patient has a fever.

But those two premises don't prove that it's malaria. There are other diseases that present with a fever, flu, for example, or Ebola. Any of these explains the symptoms, and any of these fits the premise, if this theory is true, then we will observe these symptoms. Fever underdetermines malaria, and in general a symptom underdetermines a diagnosis. To conclude that the malaria theory is true, on the basis of its successfully explaining the fever, is to commit the fallacy of affirming the consequent. It would be the same fallacy if we conclude that a theory of gravity is true, on the basis of its explaining the tides. There are plenty of alternative explanations out there.

So, the doctor, this time a good doctor, continues. What are other symptoms of malaria? If it's malaria, expect recurring episodes of shaking chills and profuse sweating. That's the prediction. And now the doctor is testing the malaria theory, not by shaping the theory to accommodate the evidence in hand, the fever, but by seeing if the theory matches evidence yet to be gathered.

Suppose the prediction turns out to be true? The patient is subsequently observed to have the episodic chills and sweating. What conclusion follows? Strictly speaking, no conclusion follows. To conclude that the malaria theory is true would be to commit exactly the same logical fallacy, affirming the consequent. The logic is the same as in the case of explanation.

(1) If the theory is true, then the phenomenon will be observed.
(2) The phenomenon is observed.
(3) Therefore, the theory is true?

No. That's the same fallacy.

But suppose the prediction turns out to be false? Is it possible to *dis*prove the theory? This is logically more promising, since the argument in this case is not a fallacy. It's a valid argument. The form is called **modus tollens**.

(1) If the theory is true, then the phenomenon will be observed.
(2) The phenomenon is *not* observed.
(3) Therefore, the theory is false.

If the patient does *not* have the predicted symptoms, at least we can rule *out* malaria. And this is what doctors do, they zero in on what the disease *is* by ruling out what it *isn't*.

Testing a theory of gravity will have the same logic as diagnosing a disease, so we should look at the predictions of the Newtonian theory and see what conclusions can be drawn.

The history of science, as usual, offers some cautionary examples against over-confidence. This time it's about over-confidence in the decisive logic of disproof. Tycho, recall, reasoned that if the Earth moves, orbits the Sun, then the angular positions of stars will change over the course of a year. No such stellar parallax was observed. He concluded that the moving-Earth theory was false. But the Earth does orbit the Sun. Similarly, Aristotle reasoned that if the Earth rotates on its axis, a dropped object would fall not straight down but some distance to the west, left behind by the moving Earth. Dropped objects are observed to fall straight down, so Aristotle concluded that the Earth is stationary. But the theory, the one that says the Earth rotates, is true, despite the false prediction.

So, the logic can't be as tidy as the malarial *modus tollens*. To understand what's going on in scientific testing, and to see how Newton's theory was tested, we'll look at two important cases, the discovery of the planet Neptune, and the non-discovery of the planet Vulcan. One was a great success for Newtonian theory, the other a failure.

Start with the success story, Neptune. Neptune is the eighth planet, both in order of discovery and in distance from the Sun. It was discovered in 1846, and the story is interesting on its own, for its quirky characters, missed opportunities, and a dramatic climax. But we are most interested in the details of the case to clarify important aspects of scientific testing. In particular, we will add to the minimal logic developed with the medical example, and point out a mutual influence between theory and observation.

The discovery of Neptune will also help clarify the distinction between the *truth* of a scientific claim and its *justification*. The limitations of science are often summarized by saying that we can never arrive at the truth (or the absolute truth, or the Truth), because we can never be certain that what we say now won't be disproved later. But this confuses the *truth* of a claim with our ability to *demonstrate* its truth. Each of our theories is, in fact, true or false, independent of our abilities to demonstrate which it is. The truth of a theory neither comes in degrees nor changes over time (unless nature itself changes). Justification, the reasons to believe that a theory is true, does come in degrees, and it changes as more evidence is available. It's not that we can never arrive at the truth, but that we can never arrive with certainty. This distinction will be clear in the case of Neptune.

No theory is tested on its own, one-on-one with the evidence. There is always a broader theoretical context to connect what you plan to test, the hypothesis, with observable expectations and results. In the case of Newton's theory of gravity, the companion laws of mechanics, the $F = ma$, link the dynamics of gravity to the observable kinematics of planetary motion. The entire theoretical package can be tested in a ready-made laboratory, the Solar system.

Together, Newtonian mechanics and Newtonian gravity are beautifully coherent. They were also remarkably successful in tests against observations of the planets. Six planets were known at the time Newton published his theory. Predictions of where each planet would move under the gravitational influence of the Sun and the

Table 8.1. *Radial positions of the planets
known in 1772*

Planet	Distance to the Sun (AU)
Mercury	0.39
Venus	0.72
Earth	1.0
Mars	1.5
Jupiter	5.2
Saturn	9.5

other planets was not a trivial calculation. It still isn't. The equations can be solved exactly only if there are only two objects present. If it were just the Sun and the Earth, the equations of motion could be solved with no trouble. But with three or more objects, there is no exact solution. The procedure is to solve the two-body problem, and then add in the perturbing effects of the other bodies. It's tedious, but ultimately precise and reliable.

Things went very well for Newton's theories, making mathematical predictions of where the planets would be, predictions that matched subsequent observations. This, of course, doesn't prove the theory is true, but it seems some good reason to think so.

The careful observations of planets paid off in other ways. In 1772, for example, the German astronomer Johann Bode revealed a mathematical regularity in the planetary distances from the Sun. Using the Earth–Sun distance as one unit, an Astronomical Unit (AU), the observed positions of all the planets known at the time are shown on Table 8.1.

Bode borrowed an idea from Johann Titus and demonstrated a pattern in these numbers. Start with the sequence of numbers 0, 3, 6, 12, 24, 48, 96, ... Except for the 0 and the 3, each number is double its predecessor. This is the regularity. Now add 4 to each number to get 4, 7, 10, 16, 28, 52, 100, 196, and so on. Divide each number by 10. The result is 0.4, 0.7, 1.0, 1.6, 2.8, 5.2, 10.0, 19.6. Except for the missing correspondent at 2.8, these numbers of Bode's law pretty well match the radial positions of the planets. The comparison is in Table 8.2.

Bode's law does not in any way explain why the planets are at these locations, but the law does make predictions. There is the gap at position 2.8. In 1801, the first of the so-called minor planets was discovered, and over time the asteroid belt filled in the space between 2.3 and 3.3 AU. Furthermore, Bode's law predicts that if there is a planet beyond Saturn, it will be roughly 19.6 AU from the Sun. In 1781, even before the gap at 2.8 was filled in, William Herschel discovered the seventh planet, Uranus, at radius 19.2 AU.

Table 8.2. *Comparison of Bode's law prediction to observed values of radial positions of the known planets. The gap at 2.8 AU was later filled in by the discovery of minor planets in an asteroid belt*

Planet	Distance to the Sun (AU)	Bode's law prediction
Mercury	0.39	0.4
Venus	0.72	0.7
Earth	1.0	1.0
Mars	1.5	1.6
		2.8
Jupiter	5.2	5.2
Saturn	9.5	10

It is noteworthy that a survey of astronomical charts made prior to the discovery of Uranus showed that the planet had been observed and recorded several times, at least as early as the first Astronomer Royal, John Flamsteed, in 1690. It had been observed, but never observed as a planet. Seeing something as a planet rather than a star requires either seeing it extended as a disk rather than a point of light, or having credible measurements of its movement relative to the fixed stars. Observation in astronomy, like observation in all of science, is more than mere physical sensation. It all depends on what you make of it.

By 1821, the orbital positions of Uranus were tabulated with sufficient precision to show an undeniable disagreement between the actual orbit and that predicted by the Newtonian laws. By 1840, the disagreement had grown to an enormous 2′ of arc, making it impossible to blame the tools or methods of observation. Various explanations were proposed, from the incidental – a comet had struck the planet and knocked it out of its Newtonian orbit – to the profound – that Newton's law of gravitation was flawed. But the judgment of the scientific establishment, for which there is no better authority than Sir George Airy, the Astronomer Royal from 1835 to 1881, was that everyone was "fully impressed with the universality of [Newton's] law of gravitation." (Quoted in Moore, 1996, p. 94.) This suggests that Newton's theory was no longer being tested; it was being used to discover new things about the Solar system. And the most likely explanation to the aberrant motion of Uranus was the effect of an exterior planet that was as yet unseen.

Sir George was not particularly anxious to find the hypothesized eighth planet. In fact, he discounted the chances of ever seeing it, saying, of the perturbation in Uranus' orbit, "If it be the effect of any unseen body, it will be nearly impossible ever to find out its place." (Quoted in Moore, 1996, p. 97.) This attitude was characteristic of Airy, a man obsessed with order and reluctant to have his tidy understanding of things, either the law of gravitation or the population of the Solar system, changed. His way of dealing with anomalies in the scheme of things was to ignore them. But then this is a man who is said to have spent the better part of a day in the basement of his observatory labeling empty boxes with the word "empty" so there would be no mistaking them.

Airy's disregard for the challenge of locating the unseen body was not shared throughout the community of astronomers. The first precise prediction of where to find the planet was in 1843 by a young English mathematician, John Couch Adams. Adams used Bode's law to estimate the planet's radial position. He zeroed in on its angular motion by doing a kind of reverse-perturbations analysis on the motion of Uranus. He sent his calculations and his results to Airy, hoping the Royal Astronomical Society would undertake to look for the planet where he told them it would be.

Airy received the calculations with dismissive skepticism. Added to his reluctance to disturb the model of the Solar system as he knew it was a general disrespect for youth. Adams, just one year beyond his graduation from Cambridge, was unlikely to have discovered a planet by pen and paper alone. Nonetheless, Airy responded by sending Adams a kind of test question, perhaps intended to gauge the young man's credibility. The question involved the deviation of Uranus from its Bode's law position of 19.6 AU, and the explanation of that deviation in terms of the hypothesized eighth planet. Adams apparently thought the question to be silly, its answer already implicit in the calculations he had sent to Airy. He never responded.

Meanwhile, others were at work on the mathematical challenge of predicting where to find the new planet, among them a French chemist-turned-mathematician, Urbain Jean-Joseph Le Verrier. Using a mathematical approach somewhat different than Adams', he arrived at a similar result in 1846. Le Verrier sent his work, including the specifications on where and when the new planet could be found, to Airy. In return, he received the same question regarding Uranus' aberration from Bode's law. Le Verrier responded. Airy now began to take the predictions of an eighth planet seriously enough to direct James Challis to have a look, using the Cambridge Observatory's 30 cm telescope.

The English astronomers had two distinct disadvantages in their search for the hypothetical planet. For one, their star charts were not the most complete, thus making it difficult to identify a novelty. For two, they were generally skeptical of the theoretical work that predicted the location. Indeed, British scientists and philosophers had a tradition of empiricism, from John Locke to Newton to David Hume. Observation is supposed to tell us what to think, not the other way around. And certainly astronomy had always been an empirical science in which observation preceded theory. So, with little respect for the precision of Adams' and Le Verrier's prediction, Challis adopted a strategy of sweeping his telescope back and forth to systematically scan the general area rather than zero in immediately on the specified spot. Weeks of searching in this patterned way failed to find the planet.

Le Verrier, meanwhile, grew impatient. On September 18, 1846, he wrote to Johann Galle of the Berlin Observatory, asking him to use their 23 cm telescope to look for the planet. Galle received the letter on September 23 and directed his observing that night to the very spot indicated by Le Verrier. Expecting to see the disk of a planet, Galle was initially disappointed. It was Galle's young assistant, Heinrich D'Arrest, who suggested comparing what they were seeing to the detailed star charts on hand, and soon they found what they were looking for. "That star is not on the map!" D'Arrest is reported to have said, and by the following night that star had moved. That star was the planet Neptune. And knowing that it was a planet they were looking at, the disk was visible.

Galle wrote to Le Verrier,

> The planet whose position you have pointed out actually exists. The same day that I received your letter, I found a star of the eighth magnitude which was not shown on the excellent chart . . . published by the Royal Academy of Berlin. The observations made on the following day determined that this was the sought-for planet.
>
> *(Quoted in Moore, 1996, p. 9.)*

Thus was Neptune discovered, right where both Adams and Le Verrier had predicted it to be.

It was embarrassing to the English astronomers, especially when they looked back at the records of their search for the planet and saw that they had in fact observed the same object. They had seen it as a star rather than a planet. It had even been recorded as worthy of rechecking, but for some reason they never did. A story goes that it was Mrs. Challis who offered her husband and a dinner guest another cup of tea before the astronomer went back to work. They accepted, and by the time they were finished, the sky had clouded over. But with more confidence in the calculations and the hypothesis predicting the planet's location, chances are he would have been less easily distracted.

The Cambridge astronomers were not the only ones to have observed and recorded Neptune without realizing that it was a planet. Galileo had seen it and called it a star. And in 1793, the French astronomer Joseph Lelande had seen it and even noted its movement, which he blamed on observational error. Neither of these people had a telescope as powerful as the 23 cm refractor of the Berlin Observatory, let alone the 30 cm refractor at Cambridge. But even with these, seeing Neptune as a planet required knowing that it was a planet. As Johann Encke, the director of the Berlin Observatory put it, "the disc can be recognized only when one knows that it exists." (Quoted in Standage, 2000, p. 122.) It's not just instrumentation that advances and refines scientific observation; theory plays both a motivational and interpretive role.

Who was the first to observe Neptune? Galileo saw it in 1612, but he didn't see it as a planet. Challis saw it before Galle, and even though the Englishman was looking for a planet, he didn't see it as a planet. Galle, with a smaller telescope but greater respect for the theory, was the first to not just see Neptune but see it for what it is, a planet. This is a clear example of theory influencing scientific observation. Adams' and Le Verrier's calculations told the astronomers where to look and what to look for. And when they looked in the right place, there was a theory in place to tell them how to interpret the images. If it's a planet it must move with respect to the fixed stars and it must, with proper magnification, be seen as an extended disk. In the case of Neptune, it was respect for the theories that made Galle look hard enough to recognize the disk.

Before there was notable difference between the predicted and observed orbits of Uranus, there was little reason to believe there was an eighth planet. Prior to Bode's law, there was really no good reason, no justification at all. By 1840 though, the irregularities in the orbit of Uranus provided some reason for believing in the existence of an eighth planet. And in 1846, when Neptune was observed, the claim was as close to being fully justified as you can get. In other words, the justification, the good reasons to believe the claim is true, increased with the accumulating evidence. But the claim was true all along. It is no truer now than it was in Galileo's or even Aristotle's day.

This makes a humble but important point. Uncertain does not mean untrue. Science does aim to deliver the truth, the absolute truth, and sometimes it succeeds. There is no hubris in saying this. What science can't deliver is certainty, that is, absolute justification. That's no reason not to aim for certainty, since even unachievable goals can keep you heading in the right direction. The responsible thing to do is to acknowledge the limitation, to keep track of the degrees of justification, and to proportion our beliefs accordingly.

The original work on comparing the orbit of Uranus to the predictions of Newtonian mechanics is another demonstration of why one failed prediction does not falsify a theory. Falsification is indecisive because it is not the theory alone that makes the prediction. When testing Newton's theory of gravity by predicting locations of planets, the reasoning involves the laws of mechanics and claims about the make-up of the Solar system as well. If the theory of gravity is true, *and* if the laws of mechanics are true, *and* if the current model of the Solar system is accurate, then this planet will be in that spot at that time. If the planet is not at the predicted spot, any one (or more) of the pieces used to make the prediction could be at fault. The hypothesis in this case is Newton's law of gravity. It made a prediction about Uranus that turned out to be false, but that did not automatically mean the hypothesis was false.

Perseverance through the challenge of falsely predicting the motion of Uranus led to a novel prediction for Newton's theory of gravity. It predicted the existence of an eighth planet, and the prediction turned out to be true. The discovery of Neptune was surely a triumph for the theory. The prediction was precise and the observation of the planet was unassailable. This does not prove the theory is true. That would be the fallacy of affirming the consequent. But it is surely some measure of success to the theory, some reason to think it's true.

If the discovery of Neptune was a success for the theory, the non-discovery of Vulcan was as much a failure. Confidence was high in both the theory of gravity and the analytic method of finding the unseen cause of an observed phenomenon. This is another inverse problem. The results are known; the challenge is to figure out what brought them about. The unusual motion of Uranus was known; Adams' and Le Verrier's accomplishment was to precisely identify the cause. It turned out to be another planet. Order was restored, and the harmony between theory and observation was renewed.

But Uranus wasn't the only planet giving Newton's theory some trouble. The orbit of Mercury, the planet closest to the Sun, was also a bit off from the theoretical predictions.

From Kepler and Newton, and our own Chapter 7, we know that planetary orbits are elliptical. Kepler had the Sun at one focus of the ellipse, but Newton moved it a bit, putting the center of mass of the Sun–planet pair at the focus. The point on the ellipse at which the planet is closest to the Sun is called the **perihelion**. If it were just these two bodies, the Sun and the one planet, Newtonian theory predicts that the orientation of the ellipse would not move against the background to the stars. The planet would complete a full orbit and return to the same spot and start over, retracing its previous orbit. But that's not what happens, because it's not just these two bodies. There are other planets in the Solar system, tugging on each other the way Neptune tugs on Uranus. This causes the ellipse to precess, so that the planet over-shoots its starting point when it completes an orbit. The orientation of the ellipse slowly rotates with each orbit. Astronomers measure this in terms of the precession of the perihelion.

Newton's theory of gravity has all these details covered, and it fully predicts the precession of perihelions. It's not at all an easy calculation, since it involves multiple objects, but it is, as the physicists say, doable. And it was done for Mercury. It was done by Le Verrier himself in 1859. His result was a predicted precession of 527 arc-seconds per century ($527''$/century) for the perihelion of Mercury. This isn't very much. An arc-second is 1/3600 of an angular degree. And it's 527 arc-seconds *per century*, not per orbit. It's a tiny amount, but well within observational precision in the mid-nineteenth century. The measured value was, and is, 565″. With Mercury, as with Uranus, the Newtonian theory of gravity made a false prediction. It was off by 38″ per century.

Letting a theory get away with a failed prediction seems to compromise the scientific standard of falsifiability. If the evidence doesn't force us to abandon the theory, what will? Le Verrier, flush from his success with Neptune, was not about to question the Newtonian theory of gravity. He knew what to do to accommodate the anomalous data, and it wasn't to change the theory. "We will certainly not be tempted into charging the law of universal gravitation with inadequacy." (Quoted in Roseveare, 1982, p. 20.) So in 1859, Le Verrier hypothesized the existence of a hidden planet, between Mercury and the Sun, responsible for the extra 38″ of precession.

Confidence in the reality of this interior planet was not nearly as high as confidence in the planet beyond Uranus, even before the observation of Neptune. With Neptune, William Herschel declared victory in a way that made actually looking for the planet almost superfluous. "We see [the planet] as Columbus saw America from the far shores of Spain. Its movements have been felt, trembling along the far-reaching line of our analysis with a certainty hardly inferior to ocular demonstrations." (Quoted in Baum and Sheehan, 1997, p. 102.) They knew Neptune was out there, even before seeing it. But this new planet, the one so close to the Sun, was dubious from the start.

For one thing, Le Verrier did not provide a precise location for the planet, as he and Adams had for Neptune. It's in there somewhere. If it is, and this is another reason for doubt, it should be visible, certainly during a solar eclipse. But it had never been seen. Perhaps it is always behind the Sun, its orbit exactly in time with our own but on the opposite side of the Sun. This is impossible by Newton's own laws. If the planet has a different orbital radius from the Earth, being much closer to the Sun, it must have a different orbital period. That's basic Kepler. Despite these obvious problems, a search for the new planet was begun.

At the same time, because of these obvious problems, alternative explanations for the anomalous 38″ were proposed. Some were in the form of additional matter affecting Mercury, but not matter concentrated as a planet. Diffuse matter, dust, if there was enough of it in the space between Mercury and the Sun, would exert a strong enough gravitational pull on the planet to perturb its precession. And this dust would also explain the so-called zodiacal light, a subtle glow in the night sky that seems to emanate from the Sun and extend along the ecliptic, growing fainter with distance from the Sun. But the brightness of the light allows for an estimate of the amount of dust, and it was much too small to provide the mass needed to explain the extra 38″ in Mercurial precession.

A compromise between the matter concentrated as a planet and diffuse as dust is a collection of asteroids, tiny not-quite-planets, in orbit between Mercury and the Sun. Again, the density of such objects sufficient to account for the observed motion of Mercury would result in the asteroids being visible.

So it was exciting when on March 26, 1859, an amateur astronomer, Edmond Modeste Lescarbault, a physician and a Frenchman at that, reported to Le Verrier that he had observed a previously unseen object transit across a portion of the Sun. This could be the predicted planet. On a visit to Lescarbault's observatory, Le Verrier was sufficiently convinced of the doctor's credibility, and the new planet was declared discovered. Le Verrier named it Vulcan, after the Roman god of fire and metal-work. It was, after all, right next to the hearth, as Kepler called the focus of the orbit.

This was not a triumph of Neptune proportions, however. The estimated mass of Lescarbault's planet was enough to provide only 1/17 of the pull required to cause the extra 38″ of precession in Mercury. And it was never seen again. Add to this an embarrassing follow-up in which Lescarbault reported finding another new celestial object, this time a star, that turned out to be Saturn. His credibility, and the alleged sighting of Vulcan, suffered.

None of the unseen-matter hypotheses worked to explain the anomalous precession of Mercury. There was simply not enough mass, based on indirect, or dubious, observations. There were also conceptual concerns. The diffuse matter would have to orbit the Sun, as would asteroids or a planet, but by Newtonian calculations the orbit would have to be tipped, not in the same plane as the orbit of Mercury, in order to affect the precession. Such an orientation, again using Newtonian theory to understanding the situation, would be unstable. Like an unbalanced washing machine, it wouldn't last.

With no new matter to blame for the failed prediction of the orbit of Mercury, question turned to the theory of gravity itself. There was no call to toss Newtonian gravity wholesale, only to tinker. Small changes were suggested, changes that would show up for planets close to the Sun, like Mercury, but not for planets far from the Sun, like Uranus and Neptune. The tinkering, in other words, would be focused on the distance parameter, the r in $F = GMm/r^2$.

Asaph Hall, in 1895, suggested a slightly revised formula for universal gravitation, $F = GMm/r^{2+\delta}$, where δ is an adjustable parameter that takes on whatever value is needed to get the formula to match the evidence. It's like the parameters in the Ptolemaic model of planetary orbits, the radius and period of the epicycle. You put in whatever value it takes to match the data, to save the phenomena. In this case it took a very small number, a very small adjustment:

$$\delta = 0.000\,000\,161\,2$$

The charitable description of this revision is in terms of theory being responsive to evidence. That's how we think science should work. But the disparaging description of the same event is of *ad hoc* tinkering with the theory, putting in that δ fudge-factor in the exponent, to accommodate the falsifying data. A theory so flexible is not falsifiable, and that's not how we think science should work. The Newtonian theory of gravity has adjustable parameters. It's not, to borrow a term from Steven Weinberg, theoretically "rigid" (Weinberg, 1992, p. 105). The general theory of relativity, we will find out, is significantly less flexible. The $1/r^2$ dependence in the relativistic formula will be a matter of principle, not a derivation from the collected evidence. It will not accommodate tinkering.

Whether in positive terms of being responsive to evidence, or negative in terms of unseemly patching theoretical leaks, the Hall theory works well under an instrumentalist attitude about science. Is the gravitational force really dependent on such an unlikely geometric factor of $1/r^{2+\delta}$? Instrumentalism makes this moot. Gravity is *as if* the dependence is $1/r^{2+\delta}$, and calculating in this way gets the predictive job done. It offers no explanation, but with this attitude, no explanation is called for. This is science, where theories are slaves to evidence, nothing more than formulaic summaries of data, and a mere summary can't, and shouldn't pretend to, explain.

Summarizing our own historical data on the attempts to reconcile theory with the anomalous precession of Mercury, no solution was perfect. Hall's theory was not only inelegant and artificial, it was also inaccurate in its predictions of the motion of the Moon. Hypotheses about unseen matter, whether diffuse or planetary, were easier to form, not having to conform to any conceptual constraints like elegance, simplicity, or coherence with other physical theories. But none of the matter hypotheses could be made to provide enough mass while remaining otherwise undetected. That much stuff would make a visible mark. In this case the absence of evidence is evidence of absence. It was clear by the start of the twentieth century that something pretty radical had to be done.

But it wasn't Mercury, or the non-discovery of Vulcan, that leveraged the radical change in the theoretical understanding of gravity. It was a more fundamental, principled flaw in Newton's theory. The change in the theory was nothing like the fix-up proposed by Hall. It was a wholesale discarding of the Newtonian theory, to be replaced by a very different way of describing gravity. The new account, relativity, will be based on principles. It's a top-down way of arriving at a theory, and it will be less adaptable to evidence and less accommodating of tinkering.

The guiding principles for the new theory of gravity are familiar, the Principle of Relativity and the Principle of Equivalence. It will be physics, not astronomy, that forces the issue. It will be the action at a distance in Newton's theory that is fundamentally at odds with the principles.

Einstein did not put together the new theory of gravity, the general theory of relativity, to accommodate the extra 38″ of precession in the orbit of Mercury. He put it together to abide by the Principle of Relativity and the Principle of Equivalence. But once assembled, the new theory in fact predicted the precession spot-on. It included the 38″. It may be more accurate to call this a post-diction rather than a prediction, since the observations had already been made. And getting a post-diction right is arguably a better test than getting a prediction right. The evidence is on the books before the theory is in the books, so there is no opportunity to interpret or select data with an eye towards hitting the theoretical target. And if, as Harold Jeffreys described the general theory of relativity, the theory "contains no arbitrary constituent capable of adjustment to suit empirical facts" (Jeffreys, 1919, p. 138), either the theory and evidence match, or they don't. This is the basis of falsifiability.

Einstein was proud that the general theory got Mercury right. Not only did the theory match the evidence, but it achieved the match in the right way, from first principles. This was unlike previous attempts invoking an unseen planet or the δ revision. These he described as "the assumption of hypotheses which have little probability, and were devised solely for this purpose." (Einstein, 1920, p. 123.) The "little probability" must stem from the obvious problem that if there was a planet we would see it. That's a problem with the content of the hypotheses. But the "devised solely for this purpose" is a problem of method. Evidence, according to Einstein, should not force the hand of theory. Theoretical change should be a matter of principle.

Back to the textbooks, then, the ones that describe gravity in Newtonian terms. The description is good enough for non-experts. It's good enough for using the phenomenon to get things done or predict how things move. But it's not good enough if you want to not only use gravity, but also want to understand gravity. It's not good enough because it's false.

9

Challenging the Newtonian Theory

The history of science, and the long-term look at the method, are often an account in which new ideas steadily build on the foundations of old. This is reinforced by Newton's famous, if somewhat disingenuous, disclaimer in a letter to Robert Hooke, "If I have seen further than others it is by standing upon the shoulders of giants" (Turnbull, Scott, and Hall, 1959, p. 416). This makes progress in science seem like a matter of adding to, and perhaps slightly revising, the good work already done. But the history of the science of gravity shows that this model of steady, cumulative scientific advancement is deceptive in its simplicity. The Newtonian theory of gravity is built on a very different foundation than the Aristotelian notions of natural motion and proper place. Some part of progress must be demolition and rebuilding from the ground up. This is even more apparent in the change from the Newtonian theory to Einstein. Newtonian gravitational theory and mechanics had to be abandoned, not simply revised or supplemented. A new mechanics, the special theory of relativity, was built upon a new foundation. The new theory of gravity, the general theory of relativity, was only then a natural addition.

Foundational principles and concepts deeply embedded in the Newtonian description of nature were found to be inconsistent with other physical theories and with important evidence. The empirical inaccuracies were apparent only in extreme circumstances, but the conceptual foundations were flawed in general. It's like the case of believing the Earth is flat. The theory matches the evidence for day-to-day and nearby measurements. The inaccuracy shows up only at very large scales, but the concept of a flat Earth is profoundly wrong. Similarly, errors in Newtonian mechanics only show up at very high speeds, but the foundational concepts are profoundly wrong at any speed. It was the special theory of relativity that pointed out precisely where things went wrong.

A discussion of the special theory of relativity is somewhat off-topic in a study of the science of gravity, but it is an indispensible detour that will eventually meet up with our main interest. It's off-topic because what makes the theory special is that it works only in some circumstances. There are only specific situations in which it can be applied, and those turn out to be exactly where there is no gravity. The special theory is *explicitly* not about gravity. But that means you have to know something about gravity to know when the theory can be used. You have to know about the cause of gravity to predict and recognize the gravity-free conditions that allow the application of the special theory.

The special theory of relativity also dodges direct discussion of gravity by being principally about kinematics. The focus is on spatial and temporal properties like length, time-duration and speed, rather than on the dynamics of force. But revisions to kinematics will require compensatory changes to dynamics, and the special theory will lead to the general theory of relativity and the new dynamics of gravity. We've seen it before. Astronomy before Kepler was a science of kinematics, and it became important evidence for the new science of gravity. The same thing will happen with the theories of relativity.

The special theory, while not directly about gravity, lays some essential groundwork for the general theory, which is. The special theory introduces the concept of **spacetime**, the four-dimensional playing field in which events take place and through which things move. It is a fundamental component of the general relativistic description of gravity. It's the thing that is curved. The special theory sets up the necessary properties of spacetime in a straightforward way without the challenging curves, that is, without the gravity. It will help to get these spacetime basics in place before turning on the gravity.

The special theory of relativity is only indirectly relevant to gravity, but it is directly relevant to the topic of scientific method. It is a clear case of revolutionary scientific change, what Thomas Kuhn called a paradigm shift. It is also a clear case of a top-down version of scientific method. It starts with theoretical principles, strictly enforced, and then derives more specific circumstantial details. Then comes the comparison to evidence. In other words, the theory is not a generalization from foundational evidence. Following this logic will bring important insight into the opportunities and challenges of this kind of principled scientific method.

The core principle at work in the special theory of relativity is, not surprisingly, the Principle of Relativity. Einstein adopted the principle from Galileo – in this case he was standing on the shoulders of a giant – and then applied it with rigorous consistency to the physics of the nineteenth century. The result was a new theory of mechanics. And when he added the Principle of Equivalence, as we will do in the following two chapters, the result was a new theory of gravity.

The Principle of Relativity, recall, simply says that the laws of nature are the same in all reference frames. This seems almost to be a tautology, true just by definition of the terms, for what is a law other than a universal generalization. It's a relation between properties that is the same everywhere. But the principle goes beyond this by saying the relation is the same not just in every place but in every situation of motion. The laws are the same whether you are moving or not; that's what made it impossible to detect the motion of the Earth. But there is a restriction. The principle applies only to *uniform* motion. In a reference frame that is speeding up or slowing down or turning a corner, you can feel this non-uniform motion, and experiments turn out differently. A ball sitting still on a table will continue to sit still, even if the table is on a moving train. But if the train suddenly slows down, or goes around a bend, the ball will roll. It will accelerate, but with no apparent force applied. This seems to violate the law $F = ma$, since we get acceleration with no force. We will have to return to this situation of accelerating reference frames, but for now we'll put it aside. We'll restrict the analysis to **inertial reference frames**, that is, reference frames that are moving with constant speed and in a straight line. In the next chapter we'll do what needs to be done to generalize the account to include non-inertial reference frames as well.

A reference frame will be associated with a coordinate system that is fixed, stationary, at some point in the frame. Consider a uniformly moving train as one reference frame, and the station it passes through as another. We can use coordinate system K (adopting the letter K as Einstein did, from the German word for coordinate) to precisely describe events relative to the train, and K′ to describe the same events relative to the station. The train moves with respect to the station, so K moves with respect to K′. Reciprocally K′ moves with respect to K, in the opposite direction with the same speed. If you take your physics class on the train, you can use exactly the same textbook as your stationary friends use in K′. That's because all the physics, all the laws, are the same in K′ as in K, as in every other uniformly moving reference frame. This is the Principle of Relativity.

It's important to be clear on what it means to say that a law is the same in different reference frames. A law, like the law of conservation of momentum, involves multiple individual properties. The value of each property may be different in different reference frames, but the form of the relations among them must stay the same. The properties may be relative, they may vary system to system, but they must co-vary, that is, change in a correlated way such that their form is the same. A law is said to be **covariant** in this way, namely, the individual properties change in a correlated way to keep the law absolute, the same in all systems. Thus, the value of momentum of an object will be different in different reference systems, depending on the relative velocity of the object with respect to each system. Momentum is a relative property. But the sum of the momenta in an isolated system of objects will not change; this is the law, and it is true in every inertial reference system.

The Principle of Relativity seems easily true for all the laws of mechanics, the laws dealing with position and motion. The law of conservation of momentum is a clear case. The laws of free-fall, as in the tower argument, are another. But there was some difficulty in the late nineteenth century with the laws of electricity and magnetism. A moving electric charge creates a magnetic field. If the charge is accelerated, oscillated back and forth, for example, both the electric and magnetic fields oscillate back and forth in the form of an electromagnetic wave. The wave propagates out from the source, the oscillating charge, and can cause other charged particles to move. The frequency of oscillation determines the specific effect. Low frequencies are radio waves. Higher frequencies are infra-red heat. Higher still are visible light, ultra-violet radiation, X-rays, and gamma rays. Regardless of the frequency, the waves all move at exactly the same speed, and this is the challenge to the Principle of Relativity. The speed of the wave, referred to simply as the speed of light, is part of the law of electromagnetic radiation. It's the actual numerical value, 3×10^8 m/s, that is in the law. You wouldn't have to measure the speed of light to know this value; you would only have to know the properties of electricity and magnetism and the laws of how they are related.

This is both profound and perplexing. No other law has the value of a speed built in. There are other constants in other laws, for example the gravitational constant G in Newtonian gravity, but the value of G is empirically determined, and, more importantly, it's not a speed. Speed is a relative property. If you are on a moving train, your speed in the reference frame of the train is zero, but in the reference frame of the station your speed is non-zero; it's the speed of the train. The value of the speed is determined only with respect to a reference frame. So we have to ask, in what reference frame is the speed of light equal to 3×10^8 m/s? And we have to answer, by the Principle of Relativity, that it has that exact value in all inertial reference frames. Light, or more generally, electromagnetic radiation, is unique. Its speed is absolute, the same in every inertial reference frame. The speed of anything else is relative.

The laws of electrodynamics cannot be covariant (the same form in all reference frames) unless the speed of light is invariant (the same value in all reference frames). It's not just that the speed of light appears to us to be the same, or is measured by us to be the same; it's that it *is* the same in all inertial reference frames. This means that the speed of light is not affected by any movement of its source. Hold a light-source in reference frame K and the light will move in K with speed 3×10^8 m/s. No surprise there. But now try to catch up to the light in a fast-moving frame K′ that is going in the same direction as the light but at, say, 2.9×10^8 m/s. The light will move in K′ at 3×10^8 m/s. That's the surprise. You can't catch up, or even make the light slow down, by speeding along with it. In any inertial system, no matter how fast or in which direction it moves, the speed of light, symbolized with the letter c, is always 3×10^8 m/s. All of the paradoxical phenomena of special relativity follow directly from this one fundamental fact, and it followed directly from the Principle of Relativity and the laws of electrodynamics.

The consequences of the speed of light being absolute are illustrated by keeping track of the fundamental kinematic properties in terms of their being relative or absolute. A property is relative if its value can be different in a different inertial reference frame. It is absolute if the value is the same in every inertial reference frame. Absolute and invariant mean the same thing.

To clarify the difference between relative and absolute properties, and to warm up with an easy example, consider the property of two events happening at the same place. Put yourself on the train and toss a ball straight up so that it comes back down and you catch it in the same hand. The toss is event A; the catch is event B. In the theory of relativity, an event is something that happens at a point in space and a point in time. A and B clearly happen at different times, but do they happen at different places, or do they happen at the same place? The answer depends on the reference frame. In the reference frame of the train K, A and B do happen at the same place. But the train itself is moving. So in the reference frame of the station K′, or anywhere fixed to the ground, B happens some distance down the tracks from A, that is, not at the same place.

Figure 9.1 shows the same two events A and B – there is just one act of tossing the ball in this example – as determined by two different reference frames K and K′. The train is moving to the right according to the station frame K′, and so B happens some distance to the right of A in K′. The train is of course not moving in its own frame K, and B happens at the same place as did A.

The property of at-the-same-place is a relative property. This should not be paradoxical at all. But now consider the property of two events occurring at the same time and ask if this property, simultaneity, is absolute or relative. Newtonian mechanics, and most intuition, say that simultaneity is an absolute, that if two events A and B happen at the same time in one reference frame then they happen at the same time in all. The special theory of relativity requires that simultaneity be relative. It is a necessary consequence of the invariance of the speed of light. The conceptual key is that the fundamental invariant of the special theory is a speed, the speed of light, and speed involves both distance and time. So if a spatial property is relative, its corresponding temporal property must be relative as well, in order for the two to co-vary in a way that keeps the speed of light invariant. If at-the-same-place is relative, then at-the-same-time must be relative as well.

Here is a particular set-up that shows in more detail the relativity of simultaneity. Get back on the moving train, but this time have the train moving to the left, just so we are ambidextrous in our ability to deal with relativity. Let A and B be two small explosions that occur at opposite ends of the train car and that send out flashes of light. A and B happen at the same time in K, the reference frame of the train. We can say this, but it's important to know how this simultaneity is actually measured. In the exact center of the car there is a device to record the arrival of each light

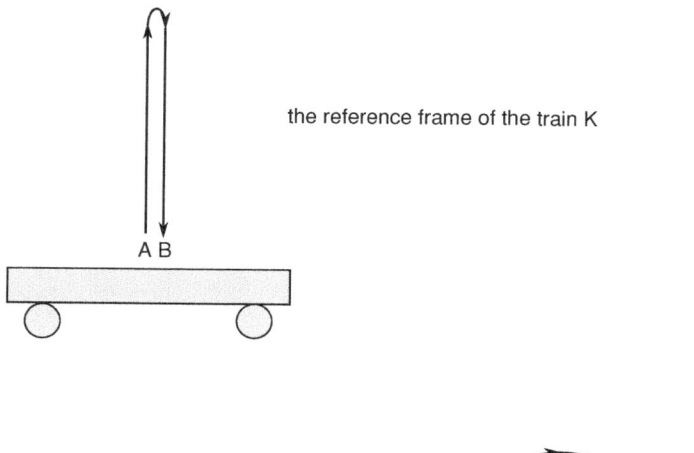

the reference frame of the train K

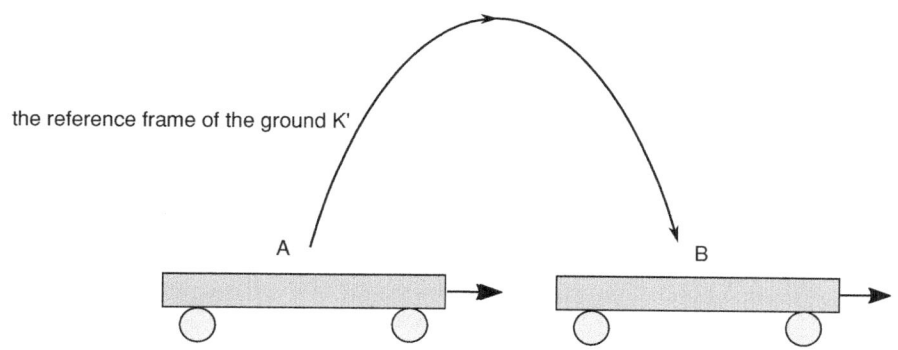

the reference frame of the ground K'

Figure 9.1. A sequence of two events A and B as determined in two different reference frames. On a train car, a ball is tossed up (this is event A) and it comes down to the same hand (this is event B). In the reference frame of the train car K, A and B happen at the same place. This is shown in the picture in the upper left. If the train car is moving to the right with respect to the ground, B happens some distance to the right of A in the reference frame of the ground K'. This is in the lower-right picture. A and B are at the same place in one reference frame but not in the other. The property of at-the-same-place is a relative property.

signal, the one from A and the one from B. It could be a wedged mirror and camera that photographs the flashes as they arrive, so there is unambiguous evidence, a photograph with the two flashes meeting at the mid-point, to show that the two events happened at the same time. Figure 9.2 shows three instants in sequence, starting with the events A and B and ending with the wave-fronts of the flashes hitting the recording device.

Here are the facts. There were two separated events A and B. Light from each arrived at the mid-point recording device at the same time. There is no ambiguity or relativity about the simultaneity at the one point in space; we have the photograph. Thus, in K, the reference frame of the train, A and B are simultaneous.

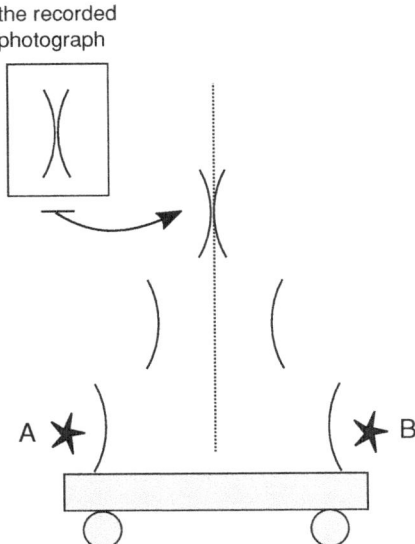

Figure 9.2. Making a record of two separated events A and B that happen simultaneously (as determined in this reference frame). A and B are flashes of light produced at separate ends of the train car. They send out wave-fronts of light that get closer together, as shown in the higher levels in the picture. The wave-fronts eventually collide and this is recorded on a photograph. Since the recording device is at the mid-point of the train car, the flashes A and B are determined to have happened at the same time. They are simultaneous.

But A and B are not simultaneous in K′, the reference frame of the ground. The only way the two flashes could arrive at the mid-point at the same time, that is, the only way we could get that photograph now in hand, is if B happened before A. To understand this, first ask what the photograph would look like if A and B were simultaneous. Figure 9.3 shows the sequence of events and flashes.

Note that this could *not* be what happened, because it results in a different photograph than the one we've got. A and B simultaneously send their flashes toward the mid-point device. But in this reference frame, the device, fixed to the train car, is moving to the left. Even though the explosions happened on the train, the light from B gets no extra speed from its moving source. The speed of light is independent of motion of the source. So, the recording device, moving toward A and away from B, will record the flash from A before it gets the flash from B. The photograph will show A but not B. Later we could take another picture showing B but not A.

So, what *did* happen to result in the one photograph with both flashes? Figure 9.4 shows the sequence of events. First B sent its flash toward the device, giving it the head-start it needed to catch up to the moving measuring device. Then A happened. Then the flashes from A and B met at the point where each strikes the device. In

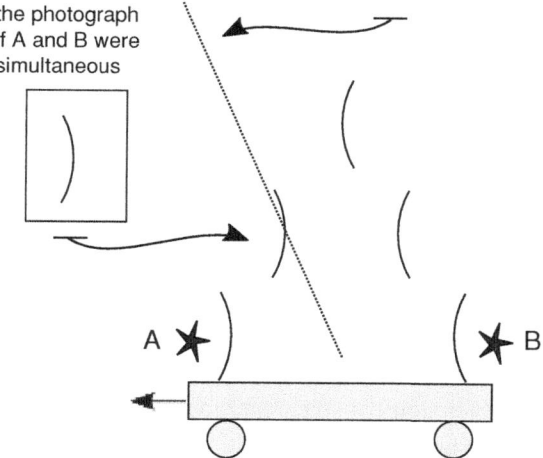

Figure 9.3. What the record would have looked like if the events A and B had been simultaneous, according to a different reference frame. This is not what happened. It results in a photograph with just one flash, and the actual photograph records both flashes. So, A and B could not have been simultaneous in this reference frame. The recording device is on the train car, so it moves to the left and hits the flash from A before the flash from B.

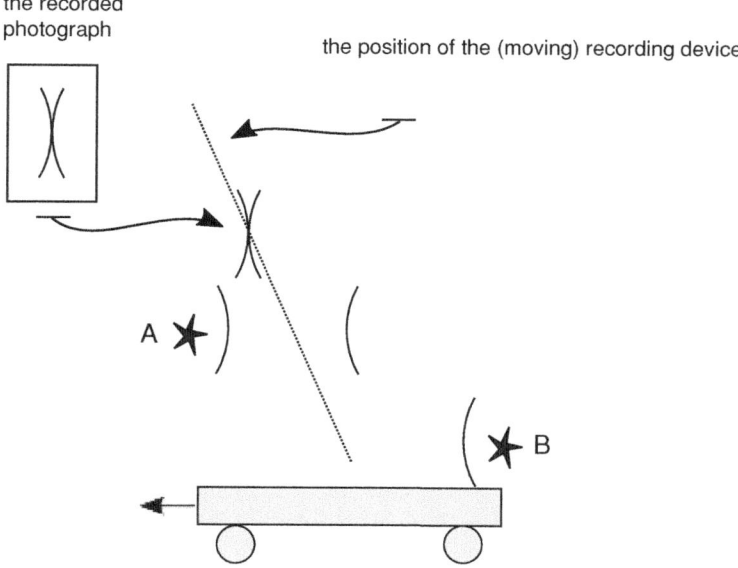

Figure 9.4. What the record indicates, according to the reference frame of the ground. B happened before A. That's the only way its wave-front could get the necessary head-start to arrive at the recording device that is moving away from B and toward A. B happened and then some time later (higher in the picture) A happened. Simultaneity is relative to reference frame.

K′, B happened before A. A and B are simultaneous in K, but they are not in K′. Simultaneity is relative.

This result is true no matter how fast the train is moving, but it would have to be moving really fast for the effect to be detectable. This is because the speed of light is really fast. If the train was going 100 km/h, the device is moving away from B but at a tiny fraction of the speed of the approaching light. This is true of all the effects of special relativity. They are always there, but they are not large enough to measure or notice unless the relative speeds are near the speed of light.

Simultaneity is relative and at-the-same-place is relative. Notice the interdependence of space and time. The value of the temporal property depends on the motion through space. Since the temporal property is not independently determined, we can no longer talk about time without also talking about space. Thus, the kinematics of relativity will have to take place not just in space but in spacetime. Each event is located in the three dimensions of space and in the one dimension of time. Spacetime is four-dimensional, but not in the sense of an extra physical, spatial dimension. The fourth dimension is just the additional parameter of an event, its location in time.

There are more consequences from the invariance of the speed of light, surprising changes to the relative or absolute status of spatial and temporal properties. Properties once thought to be absolute turn out to be relative, because a property once thought to be relative, the speed of light, turned out to be absolute.

Consider the property of time duration, simply how much time elapses between two events like the tossing and catching of the ball or the start and finish of a race. Again, Newtonian mechanics, and most intuition, say that time duration is an absolute. And again, relativity shows that it must be relative. A runner's time for the race will be different on her watch than it is on the judges'. You have to go really fast for this difference to be detectable, but it's always there. We have already demonstrated a particular case of the relativity of time duration in the case of simultaneity. The time duration Δt between the two events A and B was zero in system K – the two events were simultaneous – but non-zero in system K′ – B happened some time before A. That is, $\Delta t \neq \Delta t'$.

Here is a different thought experiment that shows how the absolute speed of light entails the relativity of time duration. Again, the details on how the properties are measured are important. So we need a clock. In this case, construct the clock with two mirrors facing each other and separated by a distance d. Send a flash of light straight up from one mirror to the other, as shown in Figure 9.5. Event A is the light leaving the bottom mirror, and event B is its return to the bottom mirror. This is the ticking of the clock. If d is a distance of one-half of a light-second, 1.5×10^8 m, then the clock ticks once per second. The time duration between A and B is one

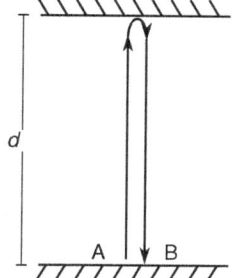

Figure 9.5. A clock that uses two mirrors and reflected light to measure time duration. The time duration Δt between events A and B, shining light from the lower mirror and its return to the lower mirror, is the distance divided by the speed of light, that is, $\Delta t = 2d/c$. If d is set to half the distance light travels in a second, the clock ticks once per second.

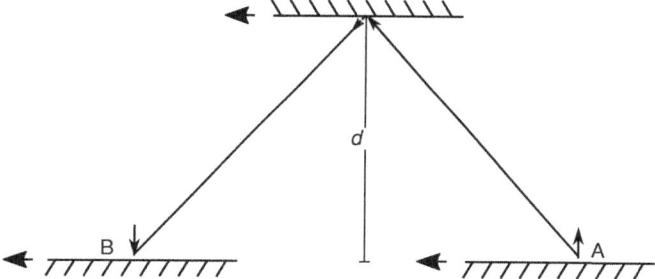

Figure 9.6. The same clock and the same two events A and B as in Figure 9.5, but now with the clock moving to the left. Again, the time duration $\Delta t'$ (primed because this is determined in a different reference frame) between events A and B is the distance divided by the speed of light. But the distance is greater than when the clock was at rest, because the A and B happen at different points in space. The distance is the hypotenuse of the triangle. Longer distance at the same speed results in a longer time. $\Delta t' > \Delta t$. Time duration is relative to reference frame.

second. More precisely, this is the time duration as measured in the rest-frame of the clock itself.

Suppose the clock is moving. Put the clock on the train and measure the time duration between the same two events but in the reference frame of the ground. The clock moves to the left with the train, so the reflected light returns to a different place, somewhere down the tracks. Figure 9.6 shows that the light has a longer path to get from A to B, the hypotenuse of the triangle rather than just the one side, longer than it had in the reference frame of the train. It has further to go but, and this is the key, it goes at exactly the same speed. So, it must take longer to get there. The time between A and B must be greater than one second, that is, $\Delta t < \Delta t'$.

Moving clocks tick more slowly. Moving people have slower heart-rates and they live longer. This is called time dilation, and the effect is symmetric. Compared with clocks at the station, clocks on the moving train tick more slowly. According to clocks on the train, clocks at the station are slow. The symmetry is a challenge to understand, but it clearly follows from the Principle of Relativity. We could have set up the clock at the station. It would then be moving (to the right) according to the train reference frame, and the light would have the longer path to follow. That would be the moving clock, according to K (the train) and it would tick more slowly.

The mathematical mechanism of the special theory of relativity allows precise comparison between Δt and $\Delta t'$. That is, if you know Δt, and you know the relative speed between K and K', you can calculate $\Delta t'$. The shift from K to K', and the formula of comparison, is called a Lorentz transformation.

It should come as no surprise to learn that if time duration is a relative property then so is distance, the spatial separation between two events. Length is relative. A train car that is 10 m long in its own reference frame will not be 10 m long in the reference frame of the station. It will be shorter. Moving objects are shorter in the direction of their motion. This is called length contraction, and again, the effect is a direct consequence of the absolute speed of light and it is symmetric.

The measurement of the length of a moving object requires two events that are either simultaneous (noting the positions of each end at the same time) or timed (noting how long it takes between the passing of one end and the passing of the other). Either way, simultaneous or timed sequence, the result will be relative to a reference frame. Here is one way to measure the length of the train car using simultaneous events. Send the train car moving to the right through a tunnel, and set things up just right so that the moving car just fits, front to back, in the tunnel. That is, the back end of the car just enters the left-hand end of the tunnel just when the front end of the car emerges from the right-hand end. The car and the tunnel are exactly the same length, call it L, according to the reference frame of the tunnel itself.

The careful relativistic analysis of the situation requires identifying specific events. The back of the car coincident with the left-hand entrance to the tunnel is event A. The front of the car at the right-hand exit from the tunnel is B. In the reference frame of the tunnel, A and B are simultaneous. But in the moving reference frame of the train car, A and B are not simultaneous. Our original account of the relativity of simultaneity shows that, in the train reference frame, B happens before A. The front of the car comes out of the tunnel before the back enters. In other words, the car does not fit in the tunnel; it's too long. In the reference frame of the train car itself, the length of the car is greater than L. In the reference frame of the ground and the tunnel, the car is moving and its length decreases, it contracts down to *L*. Moving objects are shorter in the direction of their motion.

All of these effects in special relativity are purely kinematic. There are no forces involved. Nothing pushes on the moving train car causing it to compress. The effect is entirely a result of changing reference frames. By an easy, pre-relativity analogy, if an object is to-the-left from one reference frame but to-the-right from another, we don't ask what force moved it from left to right. There is no force and it didn't move. Position is relative to reference frame. And as it turns out in the special theory of relativity, so are simultaneity, time duration, and length.

Einstein, in an essay titled "My Theory" written for *The Times* of London in 1919, reflected on the method and status of his theory of relativity. He divided the theories of physics into two kinds, constructive theories and theories of principle, tacitly allowing for two distinct theoretical methods. Constructive theories are built by generalizing on a variety of empirical phenomena, finding what they have in common and synthesizing a single, universal account. Theories of principle start with the universal law and deduce the more specific consequences. To quote Einstein, "The theory of relativity is a theory of principle." From the fundamental Principle of Relativity and the absolute speed of light, the rest follows. The method is analytic and deductive, and the consequences *must* be true if the principles are true. Empirical testing is almost unnecessary.

The effects of special relativity have been tested, though. Time dilation has been measured, as has length contraction. For example, the Stanford linear accelerator gets electrons up to nearly the speed of light by a precisely timed bait-and-switch with positive and negative charges. The separation of the bait and the timing of switch were designed to accommodate the length contraction and time dilation relative to the electron's reference frame, and it works. As it turns out, if these relativistic effects were not there the accelerator would only have to be a few inches long, instead of the two miles long it really is. This successful application of the results of special relativity is an effective test of the theory.

To say that length is relative is to say that the distance between two points in space differs from reference frame to reference frame. This is a change from what we said in Chapter 2. There, before relativity, we acknowledged that the position of each point is relative but, using the Pythagorean theorem, the distance was absolute. But now we realize that no property that is purely spatial, and no property that is purely temporal, can be invariant. Because the *speed* of light is the fundamental absolute, all absolute properties will have to combine both space and time. That means that the fundamental points to relate to each other are not simply points in space nor moments in time. They have to be events, points with both position and time, points in spacetime.

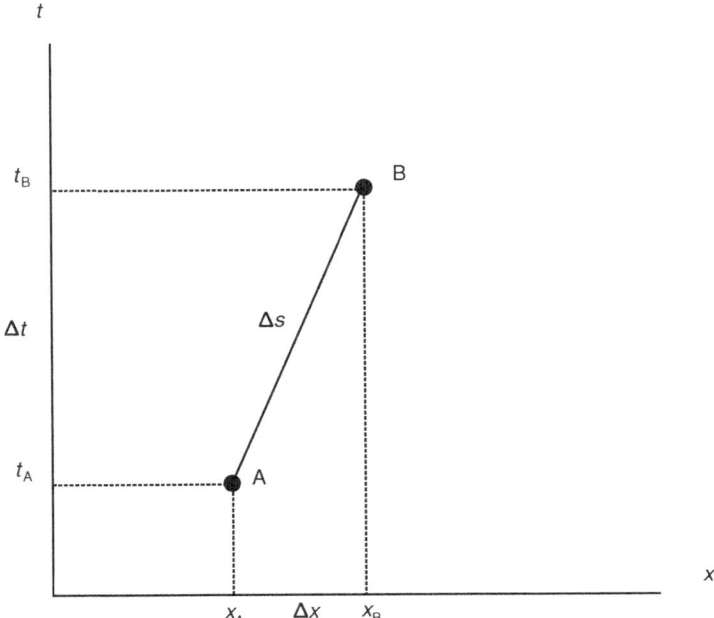

Figure 9.7. A spacetime diagram showing two events A and B and the spacetime interval between them. Only one spatial dimension, the horizontal, is shown. A is to the left of B, and A happens before B. The spatial distance Δx between A and B is relative to reference frame, as is the time duration Δt between A and B, but the spacetime interval Δs is an invariant. The spacetime interval is the same value in all reference frames.

Graphically, events in spacetime can be represented on a **spacetime diagram**. When you actually draw it on the page, the representation in very abstract, since one of the axes is time. Figure 9.7 shows a simple spacetime diagram in a particular reference frame with just one spatial axis represented. It only shows position along the horizontal, left–right, dimension. The vertical axis is the passage of time. Event A happens before B, and A is located to the left of B. A moving object would be represented by a **worldline** on a spacetime diagram, a line that is the continuous sequence of events of its being at different places at different times.

In the reference frame of Figure 9.7, we could calculate Δx, the spatial distance between A and B, but it would be a different value if we calculated it in a different reference frame. Length is relative. The same is true for Δt. Spatial separation is relative, and temporal separation is relative, but spacetime separation is absolute. In other words, there is a relativistic substitute for the Pythagorean theorem that results in an invariant spacetime distance between two events. We will develop the formula in Chapter 11, but for now it's enough to note that this property of two events, Δs on the diagram, is an invariant. It's called the **spacetime interval**.

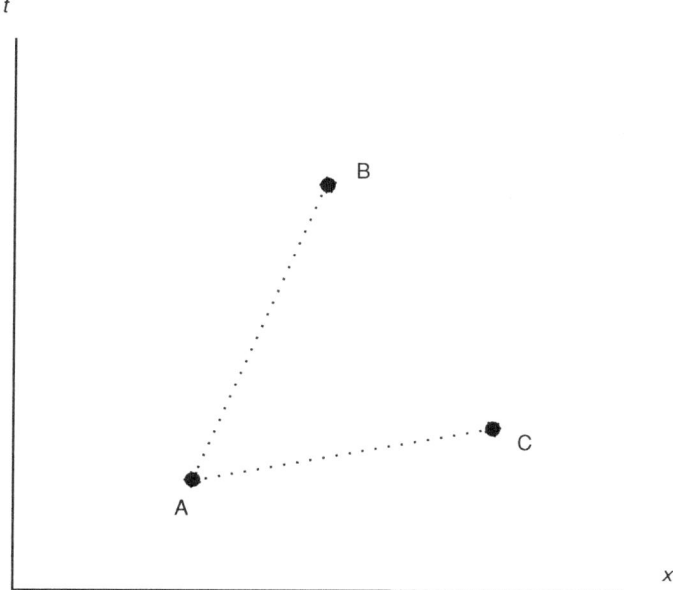

Figure 9.8. A spacetime diagram with three events A, B, and C. The spacetime interval between A and B is timelike; A can have a causal influence on B. The spacetime interval between A and C is spacelike; A cannot have any influence on what happens at C.

There is one more consequence of the invariance of the speed of light that will allow us to make effective use of the spacetime diagrams. It's a familiar one. Not only is the speed of light absolute, it is in principle an upper limit to all speeds. Nothing, at least nothing that can carry energy from a cause to an effect, can go faster than the speed of light. This can be rigorously proven, but we can get by with an informal and indirect argument. Consider trying to out-run light, that is, going faster than the light. You can try to catch up to a beam of light, but it will always be going away from you at the same speed c. As fast as you go, you can't catch up, let alone pass the light.

Apply this speed limit to the events shown on the spacetime diagram in Figure 9.8. A and B happen at nearly the same place but at significantly different times. A is the event of throwing a brick and B is a nearby window breaking a few seconds later. To get from A to B requires just a slow-moving brick. The worldline of the brick would be the line drawn from A to B. But to get from A to C, a more distant window that breaks just a very short time after the throw, requires a very fast-moving brick. Its worldline would be much less steep, since less time elapses in covering more distance. There will be pairs of events that are connected only by a faster-than-light speed. These events cannot be causally related, since there is no faster-than-light causal influence. Such events are said to have a spacelike separation. Events, like A and B, that are connected by slower-than-light speeds have timelike separation. If the connection between the two events requires exactly the speed of light, the separation is null, or lightlike. These terms are not of great importance in the development of the science of gravity. But it pays to remember that if one event is to have a physical influence on another, the two events must be either timelike or lightlike (null) separated. This aspect of the theory of relativity, like the invariance of the speed of light, will not go away when the special theory is expanded into the general.

The conceptual problem with Newtonian gravity is now apparent. What we've been calling action at a distance is not only a cause that brings about an effect without making contact; it's a cause bringing about an effect instantaneously and at some distance. That's not just a spacelike separation with a faster-than-light signal; it's a spacelike separation that requires an *infinitely fast* signal. That is physically impossible, according to the special theory of relativity. Another way to look at it is that action at a distance has the gravitational cause and effect happen simultaneously. That requires an absolute determination of simultaneity, again in violation of the special theory of relativity. The Newtonian theory of universal gravitation is inconsistent with the special theory of relativity.

This leaves two important tasks for a modern theory of gravity. The obvious one is to find a theory of gravity that is consistent with special relativity. The other is to remove the restriction on the theory of relativity, the one that allowed only inertial reference frames and ignored accelerating reference frames. Remarkably, the Principle of Equivalence is the key to doing both. The strategy is to take up the second task first, to make the special theory fully general, applicable to all reference frames, inertial and non-inertial. And in doing that, the theory of gravity shows up on its own.

10

Geometry and Equivalence

The special theory of relativity requires a distinction between inertial and non-inertial reference frames. It is restricted to the former; that's what makes it special. In the spirit of unification, it would be good to find the kinematic principles that cover *all* reference frames. Laws should be universal and apply without restriction, and just as it was progress to blend the laws of electricity with the laws of magnetism, and to unite the celestial with the terrestrial, astronomy with physics, it will be progress to develop the theory of relativity that accounts for both inertial and non-inertial reference frames, a general theory of relativity.

Recall the definition of an inertial reference frame. It's a reference frame that is moving at a constant velocity, that is, going straight at a constant speed or perhaps no speed at all. But velocity itself is a relative property, so, in the protocol of a theory of relativity we have to ask, constant velocity *with respect to what*? A train going straight means going straight along a line drawn on the ground, but this path actually bends as the Earth rotates and orbits. Unless we refer to some ethereal substance that is space itself, a reference dismissed by Mach's Principle, the constant velocity that determines an inertial frame will have to be referenced to some physical object or system of objects. The distinction between inertial and non-inertial reference frames cannot be dissolved, but it can be made conceptually and empirically precise in terms of real, massive things. It can be done true to Mach's Principle.

The strategy for generalizing the theory of relativity will be to simply eliminate the restriction to inertial reference frames and see what needs to be adjusted in the resulting description of nature. In other words, the Principle of Relativity will be revised, generalized to include all reference frames, and then strictly enforced. The consequences can then be clarified and checked. This is the same strategy Einstein used in developing the special theory of relativity. He started with the Principle of Relativity and discovered that it required the invariance of the speed of light. This in turn required revision to fundamental aspects of mechanics, revisions that have testable consequences. The general theory of relativity will develop by the same method. It, too, is a theory of principle.

To lay the foundation for the general theory of relativity, simply remove the word "inertial" from the Principle of Relativity. The laws of physics are the same in all reference frames. This is often called the **Principle of General Covariance** to distinguish the more general form from the restricted Principle of Relativity. The name is much more appropriate, since covariance means that the form of the law does not change. Individual properties are invariant if they don't change; combinations of properties are covariant if they don't change. A covariant law is an absolute law.

The Principle of General Covariance requires that no experiment can be used to detect whether the reference system is moving or not, even if the system is accelerating. This was easily true for uniform motion. In a train car with no windows, no experiment can detect a perfectly smooth, straight motion. But there seems to be an immediate counter-example to the Principle of General Covariance if the train turns or slows down. This can be detected from inside the system. If you drop a ball under these circumstances, a train-car version of the tower argument, it does not fall straight down. The law of free-fall seems not to apply to the non-inertial reference system. Remember the modern revision to the tower argument at the end of Chapter 6; the stone does *not* fall straight down if the Earth is moving. The Earth's motion *is* detectable from within the system. It's because the motion is circular, that is, non-inertial.

But this is very much like what happened in the beginning of the special theory of relativity. The Principle of Relativity ran up against what seemed like a counter-example. The speed of light was explicitly in the laws of electrodynamics, and speeds, it was thought, could not be absolute. It was coming to terms with this challenge that led to a consistent theory. It will happen again. This time the troublesome law is Newton's second law of mechanics, $F = ma$. If you put a ball on a flat, horizontal table on the train, it will sit still on the table as long as the train's velocity is unchanged. There is no force, so there is no acceleration. All is well. But if the train abruptly slows down, the ball will, with no horizontal forces applied, accelerate toward the front in the reference frame of the train. There is acceleration with no force. That's not right.

Here's a more exciting version of the experiment with the ball on the table on the train. Do it in space, outer space, in a rocket ship. This way you won't need the table. Have the ship just drifting through space with no rockets firing, in other words, going at a constant velocity. When you release the ball with no push, just gently let go, it will float exactly where you put it. This would be a perfect place to teach physics because, clearly, with no force there is no acceleration in any direction. An object at rest (in this reference frame) remains at rest, and so on. You could adjourn the class, come back the next day, and the ball would be floating in just the same spot. But if the rockets are turned on and the ship accelerates forward, the ball will spontaneously accelerate toward the back. Here again there is acceleration with no force. This experiment seems to unambiguously indicate that the reference frame is accelerating. And, since all loose objects are affected in exactly the same way, you too will fall toward the back, or at least feel the pressure of whatever prevents your fall.

All objects, regardless of size, composition, or mass, accelerate toward the rear of the ship at exactly the same rate. The acceleration of the reference system singles out a specific direction, a unique orientation, where previously all directions were identical.

A very similar thing happens all the time on the Earth. There is a unique orientation, a specific direction in which all unrestrained objects accelerate at the same rate regardless of size, composition, or mass. There is no visible force in the sense of there being springs or strings or anything in contact. The result of dropping a ball in a room on the Earth is identical to letting go of a ball in an accelerating rocket ship, if the acceleration of the ship is 9.8 m/s^2.

The point of this comparison between the physics lab on the Earth and the same experiment in a lab in space is simply that with the windows closed there would be no way to tell which lab you occupied. The effects produced by the accelerating reference frame are indistinguishable from the effects produced by gravity. There is this link between gravity and what makes a reference frame non-inertial. This is encouraging from the perspective of Mach's Principle. Since gravity is determined by real physical objects, it seems that the distinction between inertial and non-inertial reference frames will be determined by real physical objects as well.

To make good on the encouragement, note first that the experiment in the non-inertial reference frame makes it inevitable, and in this sense it explains, heavy and light objects falling at the same rate. They are just floating there while the floor, the back of the rocket ship, accelerates up to them. And since gravity produces the same result, a brick and a balloon must fall together, barring the disturbance of air. The link between an accelerating reference frame and the effects of gravity gets to the same feature of nature as the Principle of Equivalence. Galileo stated the principle in terms of heavy and light objects falling at the same rate. Newton said essentially the same thing by equating the gravitational and inertial masses of each object. Einstein named this the Principle of Equivalence and put it this way: The effects of an accelerating reference frame are indistinguishable from the effects of gravity.

Another way to understand the Principle of Equivalence is by dropping the ball, or anything else, in an inertial reference frame. It's pretty boring, and we've done it before by releasing the ball at rest in the drifting spaceship; there are no rockets firing so the reference frame is not accelerating. That's an inertial frame. The ball doesn't move; it floats without falling to the ground. That could never happen in the lab on the ground, not without some supporting force. But that's a different experiment. The only way to get the floating result on the Earth is to allow the lab itself to free-fall. Set up the lab in an elevator and cut the cable. As the lab falls toward the ground, everything inside falls with it, at exactly the same acceleration. Unrestrained objects float. In other words, a reference frame that is in free-fall in a gravitational field produces the same experimental results and observations as an inertial reference frame, like the one in the spaceship going at a constant velocity.

Two important things follow from this simple but profound fact. First, an inertial reference frame is simply a reference frame that is in free-fall in a gravitational field. The physical system, the elevator or the spaceship, has no force pushing or pulling on it, no rockets or cables. This allows the distinction between inertial and non-inertial systems to be entirely by reference to objects, the source of the gravitational field. There are details to work out, but this is progress. The second consequence, really a restatement of the first, is that the effects of gravity can be eliminated by free-fall. Phenomena in the free-falling elevator are identical to phenomena in the drifting ship in deep space. The weightless floating is evidence of no gravity, and it happens, briefly, as the elevator falls. So, you get rid of the effects of gravity in the same way you get rid of the effects of a non-inertial reference frame, by free-fall.

Put the two ideas about free-falling reference frames together, and it shows why the special theory of relativity is restricted to circumstances with no gravity. The special theory applies only to inertial reference systems, and these are the systems in free-fall. Freely falling systems eliminate the effects of gravity. Therefore, special relativity applies only where there is no gravity.

The Principle of Equivalence works to save the Principle of General Covariance from apparent counter-example, just as the absolute speed of light saved the Principle of Relativity from the challenge of electrodynamics. And just like the absolute speed of light, the equivalence of gravity and non-inertial reference frames can be tested and clarified by its particular consequences.

The first is the easiest to test. Gravity bends light. In other words, in a gravitational field a beam of light will curve inward toward the massive source, as if the mass is a lens. The effect is sometimes called **gravitational lensing**. The speed of light is unaffected, and the light is still the fastest possible signal between two points, but the trajectory, the worldline, is deflected. And being the fastest connection, it is the shortest line, that is, the straightest possible line. Put all this together and the straightest path between two points is curved by a gravitational field.

There were several steps in this derivation, and we need to work on them individually. First the part about gravity bending light. It follows directly from the Principle of Equivalence, so any theory of gravity that includes this principle will also have the consequence of gravitational lensing. In an inertial reference frame, the

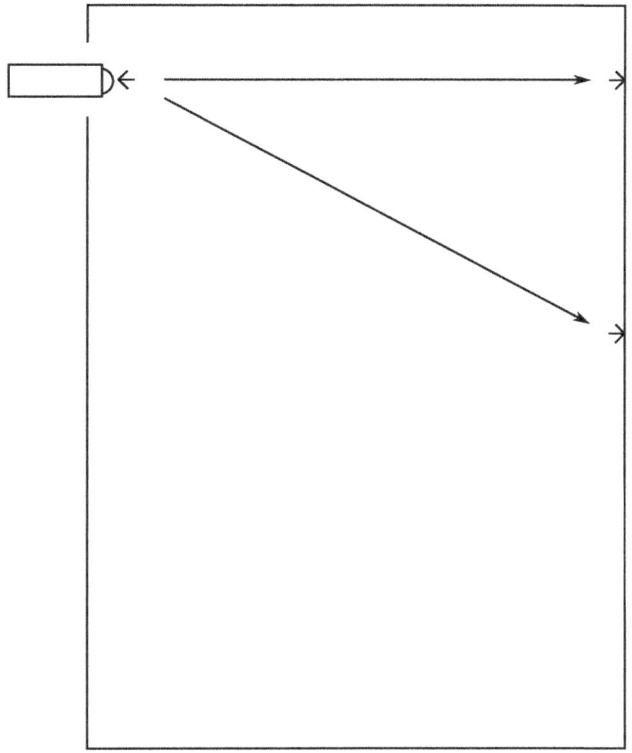

Figure 10.1. A beam of light crossing from left to right in an inertial reference frame. The beam is always straight. It will go directly across if the spaceship lab is not moving, or it will angle down if the spaceship lab is moving up with a constant speed.

drifting spaceship, for example, a light ray will follow a straight trajectory. Point a laser from one side of the lab to the spot on the wall directly opposite, and the beam will be a straight line as shown in Figure 10.1. This is just using a laser pointer to mark a spot on the screen across the room; you point and shoot and the light shows up as expected. If the lab is moving at a constant speed, that is, still an inertial reference system, the opposite wall will move up in the time it takes the beam to cross the room. The light will hit this same distance below the direct-opposite point, as if the source had been pointed down. But the beam will still be straight. This is exactly what happened in the light clock in the previous chapter, the two-mirrored system used to demonstrate time dilation. Whatever the speed of the spaceship, in the inertial reference system the trajectory of light is straight.

What happens if the rockets are turned on and the system accelerates? This is easiest to see by following the sequence of events in Figure 10.2. The left-hand side shows the lab at three different moments spaced at equal time intervals. First,

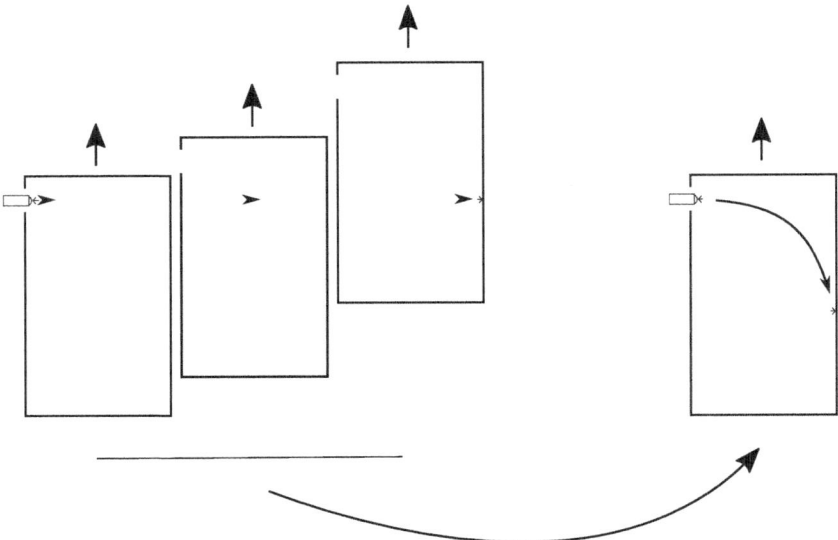

Figure 10.2. A beam of light crossing from left to right in an accelerating reference frame. On the left, three moments during the light's crossing as the spaceship lab accelerates up. On the right, the same phenomenon but with all the intermediate points filled in and as seen in the reference frame of the spaceship lab.

the light is sent from a point near the upper-left corner of the lab. Then, when the light is half-way across, the lab has moved up a little bit. Finally, by the time the light strikes the opposite wall, the lab has moved a long way up. The lab moves a greater distance during the second interval than it does during the first, because it is accelerating, speeding up. Fill in all points of the trajectory in between the start and finish, and the path of the light beam is curved down, as shown in the right-hand image in Figure 10.2. Light bends in an accelerating reference frame.

By the Principle of Equivalence, if light bends in a non-inertial reference frame then light bends in a gravitational field. Figure 10.3 simply reproduces the last frame from Figure 10.2, but it has the lab at rest on, or simply near, a very strong gravitational source, a very massive object. From inside the lab, windows shut, there would be no way to distinguish the two situations, no way to tell if the lab was accelerating or at rest in gravity.

The phenomenon of bending light, gravitational lensing, is as real and as measurable as the curved trajectory of a projectile like the bullet in Figure 3.4 and the stone in Figure 6.3. As with the effects of special relativity, the bending of light is undetectably small in our daily circumstances like accelerating up in an elevator or at rest in the gravitational field of the Earth. Since light is so fast, it gets from one side to the other before the elevator has moved much at all. It takes incredible, bone-crushing acceleration to make the curving apparent. Similarly, it takes

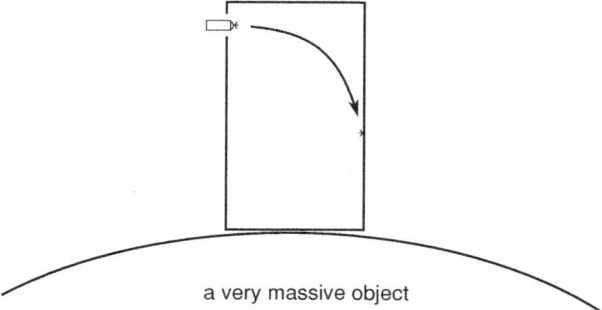

Figure 10.3. Gravitational lensing. The trajectory of a beam of light in an accelerating reference frame (Figure 10.2) is indistinguishable from the trajectory of light in a gravitational field. Here the light is bent by the massive object creating the gravitational field.

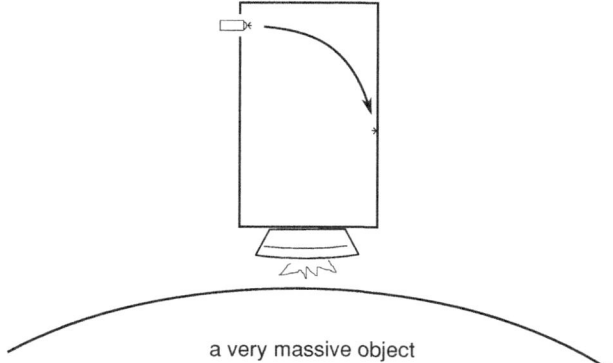

Figure 10.4. A beam of light crossing from right to left in a lab held in place near a massive object. This is exactly the same situation as in Figure 10.3 except it uses rockets rather than the ground to support the lab. The massive object is still there, and it will cause the beam of light to bend.

enormous, astronomical gravitational strength to cause detectable lensing. The effect is always there, just too small to detect except in extreme circumstances.

The prediction of gravitational lensing will be put to a test, not in a lab on the Earth, but by finding an appropriate situation in outer space, some very massive object and a traceable ray of light. To prepare for the actual measurement, first consider this thought-experiment. Position a physics lab very close to a very massive object, like a star or a huge planet, as shown in Figure 10.4. To keep the lab stationary, rockets will have to be constantly firing to counteract the gravitational pull of the star or planet. There are windows in this lab, so a ray of light can shine in through one side and out the other. If light comes in from the left and crosses the lab, the ray of light will bend down. Here again, it's just the image from Figures 10.2 and 10.3 reproduced.

The presence (or absence) of the lab has nothing to do with the behavior of the light. The reference frame doesn't cause the light to bend; the massive star or planet does that. Remove the lab and the light still bends. This is the essence of gravitational lensing, a consequence of the Principle of Equivalence, and it's one way to test a theory of gravity. All theories that include the Principle of Equivalence, and this includes both Newtonian gravity and the general theory of relativity, predict gravitational lensing, but different theories predict different amounts. The testing will have to be quantitative and precise to distinguish one theory from another.

Continuing the thought experiment, put the lab back in place with the light passing through. Now turn off the rockets and let the lab free-fall toward the star. Rays of light passing through will now be perfectly straight. With the rockets off, this is an inertial reference frame. In free-fall there is no gravity and hence no lensing. These are the circumstances in which the special theory of relativity applies.

It's worth pointing out that the inertial reference frame cannot extend throughout the entire freely falling lab. This is because the gravitational field is inhomogeneous. The center of the lab follows the field line toward the center of the star, but the edges do not. Those field lines angle inward, toward the center of the lab, and floating objects would be gently drawn inward. Furthermore, the field is stronger at the bottom of the lab than at the top, so floating objects would fall faster below than above. These are tidal effects, and they distort from the inertial homogeneity of the reference frame. The moral of the story is that an inertial reference frame is a very localized system. There cannot be an inertial frame that extends globally, throughout the universe or any extended space that is punctuated by massive objects.

Gravity affects the passage of light through space. It also affects the passage of light through time. This is to be expected, given the blending of space and time in the theory of relativity. It's not the speed of light that is altered; this time it's the frequency. The effect is called the **gravitational red shift**, since the gravitational field will have the effect of lowering the frequency of any electromagnetic radiation. If the radiation is in the visible-light part of the spectrum, lower frequency means a shift toward the color red. The amount of shift, and the actual resulting frequency, depend on several factors, and not every wave presents as red, so the name is a bit misleading. The more general description of the phenomenon is that the frequency of any electromagnetic wave will be lower the further down it is in a gravitational field. The field defines the directions of up and down. The periodic oscillations of the wave, and any other periodic process, slow down closer to the source.

The Principle of Equivalence shows that this must be the case. Put two identical clocks in an accelerating lab such as the spaceship with rockets firing. Put one

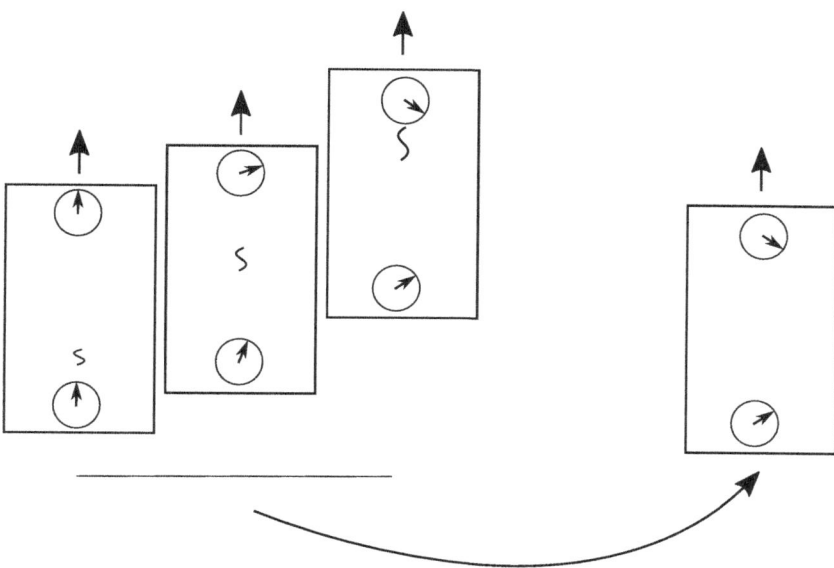

Figure 10.5. Identical clocks at the top and bottom of an accelerating reference frame. The bottom clock sends a light signal to the top clock, with the frequency of the light matched to the ticking of the bottom clock. On the left, three moments during the light's trip to the top clock. When the light arrives, the top clock is moving faster than was the bottom clock when the light was emitted, so the light is Doppler-shifted to a lower frequency. The bottom clock ticks at a lower frequency relative to the top clock; the bottom clock is slower. On the right, the relative rates of clocks happen even with no light sent.

clock on the bottom of the lab and the other at the top, where top and bottom are determined by the direction of the acceleration. This is shown in Figure 10.5. Using a beam of light that travels from bottom to top, we will show that the clock at the bottom runs slower than the clock at the top. This is the gravitational red shift. To compare the rates of the two separated clocks, synchronize the bottom clock to the frequency of the light to be beamed up. As the light signal passes the top clock, measure the frequency. Since the lab is accelerating, in the time it takes for the light to get from one clock to the other, the top clock has picked up speed. This means that the clock receiving the light at the top is moving away from the clock that sent the light from the bottom. Relative motion between the sending and receiving of a wave causes a **Doppler shift**. If the two are moving toward each other, the shift is to higher frequency. Moving away shifts to lower frequency. That's what causes the sound of a passing train horn to drop in pitch as it passes by, changing from approach (moving-toward, higher frequency) to leaving (moving-away, lower frequency). Back in the accelerating lab, the frequency of the light, as measured at the top, is lower than it was at the bottom, since the top clock is moving away from the source. The top clock will tick faster than the bottom, since the bottom was synchronized to the frequency of the light.

What happens in an accelerating reference frame happens in a gravitational field. Simply replace the rockets with a massive gravitational source and the directions of up and down are determined by proximity to the mass. The frequency of light decreases as the light climbs up the field. This general relativistic effect is not symmetric in the way that the special relativistic time dilation is. Time dilation occurs when two clocks have different velocities, and velocity is a thoroughly relative property. There's no sense to saying one clock is moving and the other is not, at least no sense without specifying a reference frame. But gravitational red shift occurs when two clocks have different positions in a gravitational field, that is, different positions with respect to the physical source of gravity. This is not relative. One clock is further from the source, and that clock runs faster than the one that is closer.

All of the analysis of the Principle of Equivalence so far has been in terms of the behavior of light. Just as in the case of the special theory, we need to add the implications for physical objects and events as well. We started with light because it is the fastest connection between two points, and hence a way to determine the straightest possible line. But now the straightest possible line is being bent. There are more familiar situations in which this happens, and they can be used to make sure the relativistic situation makes sense.

Between two points on the spherical Earth, San Francisco and Geneva, for example, there is a unique path that is the shortest distance, the straightest line. It's the route taken by pilots in the interest of minimizing both time and fuel costs. It's the distance "as a crow flies." And there is an important and geometrically well-defined sense in which this straightest possible line is curved. The curvature shows up most explicitly in geometric figures. Connect three points on the surface of the globe using straight lines. That is, draw a triangle. Now measure the interior angles of the triangle, add the three measurements together and the result is not 180°. In other words, these lines do not abide by this basic rule of Euclidean geometry, and this is because they bend. One triangle in particular shows this in a big way. Start at the north pole and draw a straight line down to the equator. This follows a line of longitude. Turn left at the equator and go straight along the equator for one quarter of the way around the Earth. Turn left again and go straight back up to the north pole. This triangle is shown in Figure 10.6. Each of the three interior angles is a right angle, so the sum is 270°.

The straightest possible line between two points is called a **geodesic**. On the two-dimensional surface of a sphere like the Earth, the geodesics are segments of the great circles. All lines of longitude are great circles, but the equator is the only line of latitude that is a great circle. To go straight from one point on the equator

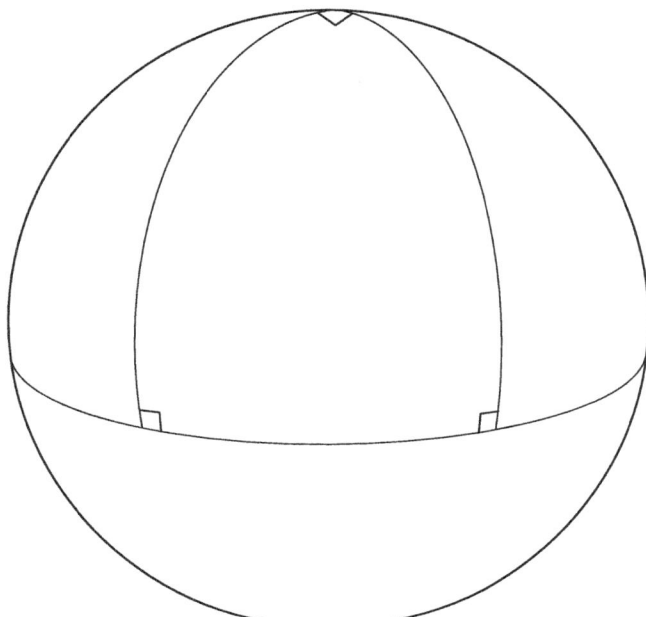

Figure 10.6. A triangle drawn on a curved surface. On the two-dimensional spherical surface, the rules of Euclidean geometry do not apply. In the triangle shown, the sum of the interior angles is 270°.

to another, to fly from Nairobi to Quito, say, you would actually move along the equator. But to get straight from Vancouver to Prague, both at about 50° north, you would fly north over Greenland, since the 50° latitude line is a longer route. The geodesic goes over Greenland. Another way to think about this, and to actually measure it, is with a globe and a piece of string. Hold the string at the two points and pull it taut. The taut string will lie on the geodesic. If you lay the string on the 50° latitude line between Vancouver and Prague, and then pull it taut, it will slide up, over Greenland, to minimize the distance. The minimum distance is the taut-string distance. The straightest possible path, the geodesic, is along the taut string.

Geodesics on the curved two-dimensional surface of the globe do not follow the same rules of geometry as geodesics on a flat surface like a table-top or a blackboard. The geometric properties of figures like triangles, circles, and parallel lines are different on a curved surface than on a flat surface. They are non-Euclidean. And by making geometric measurements you can tell whether the surface you are working on is flat (Euclidean) or curved (non-Euclidean). You could determine that the Earth is round, that it is a curved surface and just how much it is curved, by laying out triangles drawn using geodesics, taut strings, and measuring and adding up the interior angles. If the sum is 180°, the surface is flat. If the sum is anything other than 180°, the surface is curved.

On the two-dimensional surface of a sphere, the geometry is fully determined within the two dimensions. The geometry is intrinsic to the surface, without reference to any external perspective. The curvature is **intrinsic curvature**. It does not require a higher-dimensional space in which the surface is situated and into which it curves. A sphere has intrinsic curvature, but a cylinder does not, even though it is embedded in three-dimensional space. If you construct a triangle on a cylinder and measure the three interior angles, the sum will be exactly 180°, regardless of the size, shape, or orientation of the triangle. In the sense of intrinsic geometry, a cylinder is flat; its intrinsic geometry is Euclidean.

You can make a cylinder out of a flat piece of paper just by bringing two opposite edges together. You don't have to stretch the paper at all, and this is the key to intrinsic geometry, stretching. You can't make a sphere out of a flat piece of paper; you need to start with a piece of *rubber* sheet. Making a sphere will require stretching the sheet as well as bringing the edges together, and it's the stretching that affects the geometry. Draw a triangle on the flat Euclidean sheet, paper or rubber, and the interior angles sum to 180°. Roll the sheet into a cylinder and no line is deformed or stretched, so the angles still sum to 180°. But form the sheet, and this time it has to be rubber, into a sphere and the lines bow and stretch. This is what changes the intrinsic geometry. For this reason, it is helpful to think that when a surface or higher-dimensional space is described as being curved it's not so much curvature of the space as it is stretching.

Geodesics, the straightest possible line between two points, bend when they are on a curved, stretched, surface. Light follows geodesics, and gravity causes light to bend. The logic compels us to say that gravity is associated in some way with the curvature of the space through which light travels. Relativity compels us to upgrade this to curvature of the spacetime through which light travels.

Another example of geometry on the spherical surface of the Earth helps to connect the behavior of geodesics to gravity. Draw two short lines that are both perpendicular to the equator, as shown in Figure 10.7. These line segments are parallel, since two lines that intersect a third at equal angles are parallel. But now extend the lines in both north and south directions. Keeping the lines straight means following lines of longitude, and so the lines meet at both north and south poles. Parallel lines intersect, or you might say there are no parallel lines on this surface. Either way, this is contrary to the so-called parallel postulate of Euclidean geometry. It's another intrinsic feature of the surface that shows its curvature.

The two lines in Figure 10.7 are geodesics that bend toward each other. This is exactly what happens when two rays of light pass on opposite sides of a massive object like a star, large planet, or black hole. Cut and paste the two geodesics from

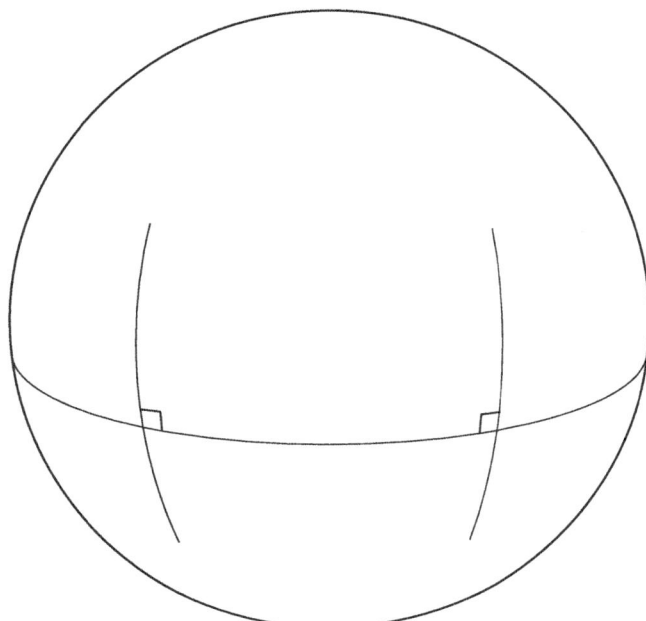

Figure 10.7. Parallel lines intersect on the curved surface. The two lines of longitude are parallel where they cross the equator, but they intersect at both poles.

Figure 10.7 onto Figure 10.8, and see them as the paths of light. This is gravitational lensing, bending the geodesics. Light from the very distant source at the bottom of the figure is focused by the gravitational field of the mass in the middle. The mass bends, or stretches, the intrinsic geometry, and this is revealed by the geodesics, the trajectory of the light.

It's worth doing just one more example with intrinsic curvature before adding in the details of how gravity is linked to curvature. Draw a circle around the north pole. That is, use the north pole as the center and lay out the points equidistant from the center. Measure the diameter and circumference of the circle. If the circle is small, the ratio of the circumference to diameter will be approximately π, as specified by Euclidean geometry. But as the circle gets larger, the diameter increases faster than the circumference. This is how the sheet had to be stretched to make the sphere. For a large circle, the ratio of circumference to diameter will be some number less than π. A specific example would be the circle centered at the north pole with a radius down to the equator, as shown in Figure 10.9. The radius is ¼ of the distance around the sphere, and the circumference is exactly the distance around the sphere. So the ratio of circumference to diameter is 2, not π.

On any curved surface, or in any curved space, small-scale measurements will be Euclidean. The non-Euclidean features show up only in larger, global figures. The

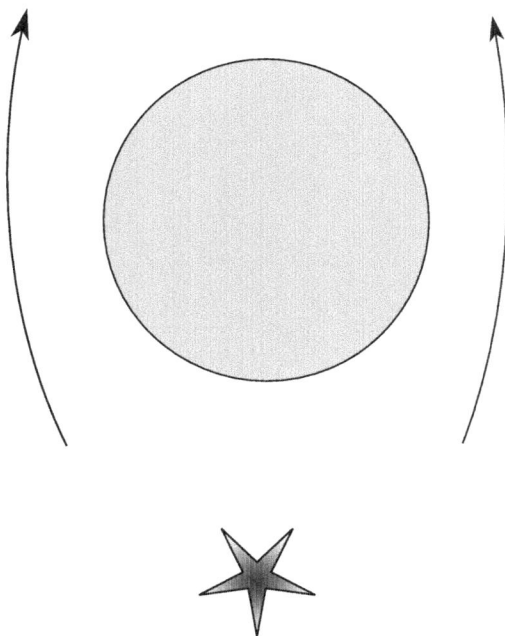

Figure 10.8. Gravitational lensing can cause parallel lines to intersect. The two light beams are parallel when they are on exact opposite sides of the massive object, but they intersect when they are some distance away, off the top of the picture.

space will be locally flat and Euclidean, but globally curved and non-Euclidean. This same distinction applies to inertial and non-inertial reference frames. A reference frame can be only locally inertial; there are no extensive, global inertial reference frames. The correlation is perfect. Inertial reference frames have no intrinsic curvature. They are flat and Euclidean. Geodesics are straight and true to the laws of Euclidean geometry.

Curved geodesics and non-Euclidean geometry seem a long conceptual way from the Principle of Equivalence. It's worth retracing the steps that got us to curved space and, in the next chapter, curved spacetime.

The Principle of Equivalence was introduced to save the Principle of General Covariance from an apparent violation of the law of inertia. That's the law that says if there is no force on an object it will not accelerate. In the language of a spacetime diagram, "not accelerate" means follows a straight worldline. A free particle, an object with no force applied, will follow a straight worldline. A straight worldline is a geodesic. We have developed the concept of a geodesic in terms of the trajectory of light, but it applies also to objects like stones and planets. There is a precise determination of their going straight from A to B, and this means following

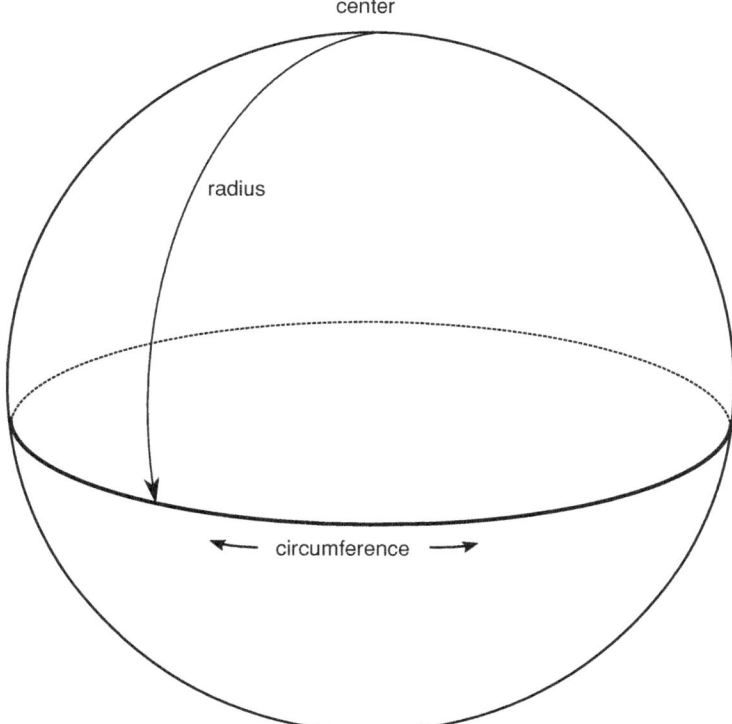

Figure 10.9. A circle drawn on a curved surface. This is another example to show that, on the two-dimensional spherical surface, the rules of Euclidean geometry do not apply. In the circle shown, with the center of the circle at the north pole and the radius reaching all the way down to the equator, the ratio of the circumference to the diameter is 2. On a flat Euclidean surface the ratio would be π.

the geodesic. So the law of inertia can be generalized to say that a free particle will follow a geodesic in spacetime. But in a non-inertial reference frame, the trajectory of a free particle curves; the particle accelerates. The worldline is curved but it is still a geodesic, because the geodesic itself is curved. The law of inertia is still true, even in the non-inertial reference frame.

There is still work to be done filling in the details of this argument, and it's the work to be done in the next chapter. That's where the actual theory of gravity, the general theory of relativity, will be made explicit. But at this point there are some fundamental features of the theory in place, requirements that the actual theory will have to meet. For one, at any point in space or spacetime, there being gravity or not is dependent on the reference frame. That is, gravity can be, in the language of relativity, transformed away. Wherever you are and whatever large object is, or isn't, nearby, there will be a reference frame in which the effects of gravity disappear. Gravity can be zeroed out by transforming to a freely falling reference frame. This means that gravity cannot be described as a force applied in an inertial reference frame, as it is in the Newtonian theory. An inertial reference frame is in free-fall, and that means there is no force of gravity. The new theory of gravity will have to be radically different and not just a revision in describing the force of gravity. Gravity is not a force at all.

The new theory will have to allow that gravity is what determines the difference between an inertial and non-inertial reference system. The new theory must also be in terms of the link between gravity and the shape, the curvature, of geodesics. It has to be a geometric theory, or, again in the language of relativity, a metric theory. The gravitational field will not be a force field, since gravity is not a force, but a metric field, indicating how the measure of shortest interval is a function of space and time. This is the challenge for the general theory of relativity.

11

The General Theory of Relativity

It will help to start with a summary of the results of the previous chapter. The link between gravity and the intrinsic geometry of four-dimensional spacetime is the conceptual groundwork for the general theory of relativity, and the little bit of math we do in this chapter will not quite make sense without a clear understanding of the important conceptual accomplishments of the last. The goal for this chapter is to clarify the general theory of relativity and to use it to answer the basic question, "Why does the dropped object fall?" But first, the review.

Light follows geodesics. That is, the path of a ray of light is always the straightest possible path through space or spacetime. It follows what we have called the taut-string line. The path of a ray of light will be curved in an accelerating, non-inertial reference frame, and so, by the Principle of Equivalence, it will be curved by gravity. This is what it means to identify gravity with geometry, since curved geodesics are a feature of a curved surface. The geodesics on the curved surface of the Earth, the great circles, result in geometric formulae unlike those of Euclid. The geometry of a curved surface is non-Euclidean. Gravity makes the geometry of the space through which light passes, the four-dimensional spacetime, non-Euclidean.

Gravity as geometry is often explained by the rubber-sheet analogy. A bowling ball on a tight rubber sheet bends the fabric in a way that a marble shot nearby will spiral into the well. The heavy object seems to attract the lighter one. There are flaws in the analogy, but one feature is spot on: the distortion of the geometry and the resulting curved trajectory of the marble are the result of actual stretching of the fabric. It's not the bending of the two-dimensional surface into another third dimension. The geometry is intrinsic to the two dimensions, and measurements of geodesics restricted to the rubber sheet itself would reveal non-Euclidean results. Simply bending the surface does not alter the intrinsic geometry at all. A cylinder has no intrinsic curvature, since its formation requires no stretching. Measurements with geodesics on a cylinder will yield all Euclidean results. To get a different geometry, there has to be some stretching. Gravity is not a matter of bending spacetime; it's a matter of stretching spacetime. That's how the geodesics are curved.

Gravity as geometry also allows us to determine the difference between inertial and non-inertial reference frames in a way that does not depend on or refer to an absolute space. The difference is in the shape of the geodesics. An inertial reference frame has straight geodesics. Three-dimensional spatial measurements result in Euclidean geometry. Four-dimensional spacetime measurements result in the special theory of relativity four-dimensional geometry called Minkowskian geometry. A non-inertial reference frame has curved geodesics. Three-dimensional spatial measurements result in non-Euclidean geometry. Four-dimensional spacetime measurements result in non-Minkowskian geometry. And since gravity determines the shape of geodesics, and massive objects are the source of gravity, massive objects determine the inertial frames. A truly inertial reference frame is one in free-fall toward a massive object. In a free-falling elevator, light rays go straight. It's an inertial reference frame. There is no universe-wide inertial reference frame, since each local massive object determines its own local inertial reference frame. The global geometry of spacetime is a patchwork of local inertial reference frames.

Gravity as geometry requires that the formal, mathematical, description of gravity be a metric theory. It has to be in terms of massive objects affecting geodesics, that is, distances between points, and then other massive objects moving in reference to the geodesics. The geodesic is the key. The gravitational field must be a metric field.

The general theory of relativity is a metric theory. To make the theory precise, a requirement for both understanding and testing the theory, we need a clear connection between geometry and the metric, the measure of distance. Geometry is about shapes, things like triangles and circles, and those are determined by distance. A circle, for example, is the set of points all the same distance from the center. An equilateral triangle is composed of three sides, all the same length, the same distance. The key to describing geometry is describing the distance between points, that is, the length of the interval between two points.

Start simple, with two points in space, A and B. To get to relativity we will have to promote these to being two events in spacetime, but to clarify the basic idea of the interval and the metric, points in space will do. The distance between A and B is the length of the interval *l*. This simple situation is shown in Figure 11.1.

The positions of A and B are most effectively described by using a coordinate system. The two points are in fact where they are independent of any coordinate system, but they are best *described* by reference to a coordinate system. The natural choice in this case is a Cartesian coordinate system with the *x* axis horizontal and the *y* axis vertical. But this is not the only choice and, just as in the development of the special theory of relativity, we will want to keep an eye on which properties depend

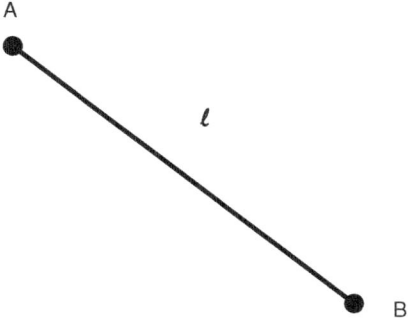

Figure 11.1. Two points in space, A and B, and the distance between them, l.

on the choice of coordinate system and which do not, that is, which properties have different values in different coordinate systems and which have the same value in all reference frames.

We can use the chosen reference frame to give a precise value for the positions of A and B, and the value of the length l. As Figure 11.2 shows, the length l is just the result of the Pythagorean theorem. Figure 11.2 is exactly the same physical situation as depicted in Figure 11.1, simply with the addition of the chosen coordinate system to facilitate precise description and application of some mathematics.

By the Pythagorean theorem,

$$l^2 = (x_B - x_A)^2 + (y_B - y_A)^2 \tag{11.1}$$

$$l^2 = (\Delta x)^2 + (\Delta y)^2 \tag{11.2}$$

Here, as in previous chapters, the Δ indicates the change in the value of the property.

Choosing a different coordinate system to describe the same physical situation will result in different values for the coordinate positions of A and B, and different values for the changes in coordinate positions Δx and Δy, but the length, the distance from A to B, will be exactly the same. Choose a different coordinate system as shown in Figure 11.3, for example, and clearly $x'_A \neq x_A$, and $\Delta x' \neq \Delta x$. But if we do the math we'll find that $(\Delta x')^2 + (\Delta y')^2 = (\Delta x)^2 + (\Delta y)^2$. That is, $l'^2 = l^2$. The length is invariant.

If you enjoy this sort of algebraic activity, you might apply one more coordinate system to describe the length between A and B, one in which the axes are not perpendicular, as in Figure 11.4. On the other hand, if you find the algebra off-putting you may skip this paragraph and Figure 11.4 entirely, with no harm done to the subsequent understanding of the general theory of relativity. But with the axes skewed, the Pythagorean theorem no longer applies, since the resulting triangle is

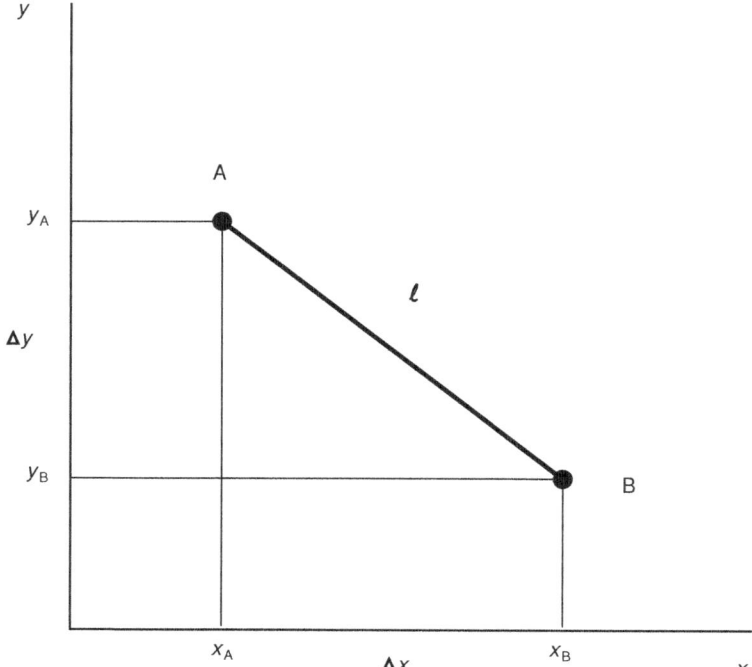

Figure 11.2. The same two points as in Figure 11.1, this time with a coordinate system. A reference frame and coordinate system are required to describe the positions of A and B. The length l is the hypotenuse of a right-angled triangle with sides Δx and Δy. Hence l can be calculated using the Pythagorean theorem.

not a right-angled triangle. Trigonometry, the more general study of all kinds of triangles, provides the formula for the length l:

$$l^2 = (\Delta x)^2 + (\Delta y)^2 - 2\cos\alpha\,\Delta x\Delta y \qquad (11.3)$$

The third term is called a cross-term, since it involves a multiplication of intervals $\Delta x\Delta y$ along both axes. It appears because of the skewed axes. But again, use this formula in this coordinate system and the value of l will be exactly the same as it was in the previous two coordinate systems, or indeed any coordinate system. Note that l is invariant.

The Pythagorean theorem is the formula to use to find the invariant length between two points A and B using their coordinate positions, but only if the two points A and B lie on a flat surface (and the coordinate axes are perpendicular). The Pythagorean theorem is a formula of Euclidean geometry. If the two points lie on a curved surface, like two cities on the curved surface of the Earth, this formula will not work. Nor will it work when we consider A and B to be events in the

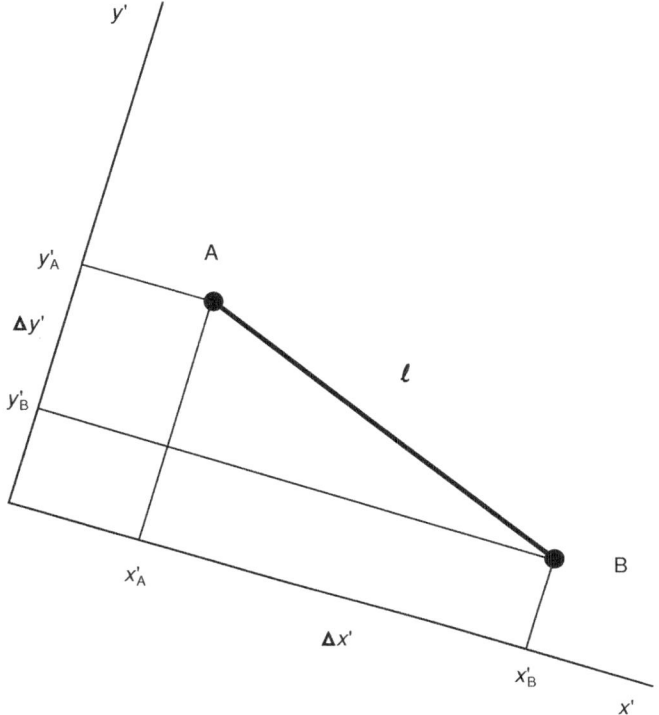

Figure 11.3. The same two points as in both Figures 11.1 and 11.2, this time with a different coordinate system. The values of positions are different in this coordinate system, but the length *l* is the same. The length is invariant.

four-dimensional spacetime of relativity. The Pythagorean theorem does not factor in the coordinate dimension of time. Nor does it respect the fact that the speed of light is invariant. The distance formula will have to be modified to account for both of these features, curvature and the absolute speed of light, to function in a relativistic theory of gravity.

We'll make the modifications one step at a time, starting with curvature, and then adding the considerations of time and the speed of light. On a curved surface, for example the two-dimensional spherical surface of the Earth, the distance between two points will still be independent of the choice of coordinate system. The distance between New York and Paris is the same whether you center your coordinate system at New York or at Paris, or at Cairo, for that matter. The interval, to use the more general term, is still invariant. But the formula for calculating the interval, given the coordinate positions of the two points, New York and Paris, will not be the Pythagorean theorem. This is because the Earth is not flat. The specific formula to use will depend on the specifics of curvature of the surface. This is good,

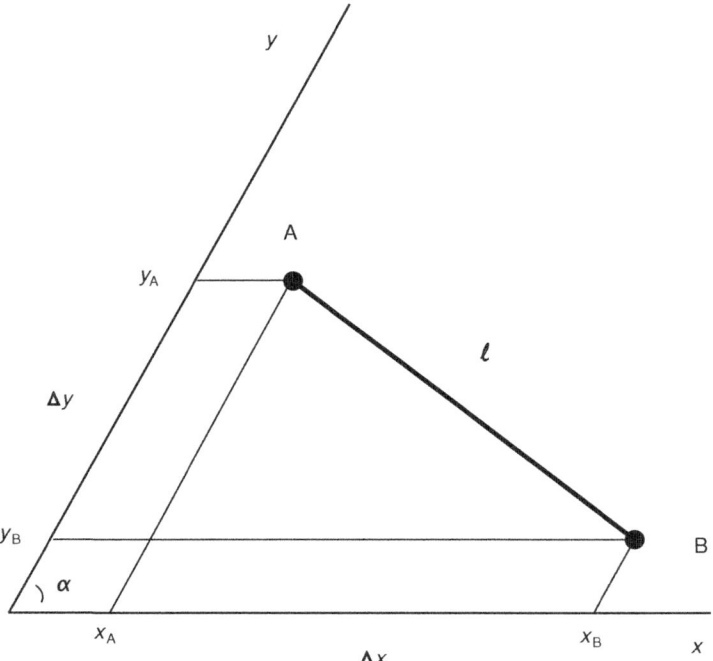

Figure 11.4. The same two points as in Figures 11.1, 11.2, and 11.3, this time with a coordinate system that has non-perpendicular axes. The values of positions are different in this coordinate system, and the formula for calculating the distance l is not the Pythagorean theorem, but the result, the value of l, is the same. The length is invariant.

since it means that if we are confined to the surface itself, we can use the details of whatever formula for calculating distance to tell us exactly how the surface is curved. It's intrinsic curvature, don't forget. It's determined and determinable on the surface itself. This will still be true when we extend the geometric analysis to four-dimensional spacetime.

The formula for calculating l^2 is called the **metric**. The Pythagorean theorem is the metric of a flat, that is, Euclidean, surface. Adding a factor for $(\Delta z)^2$ and you get the metric for flat, Euclidean, three-dimensional space: $l^2 = (\Delta x)^2 + (\Delta y)^2 + (\Delta z)^2$. If you are using a coordinate system with perpendicular axes, and you find that measured distances correspond to the calculated results of this metric, then you know you are on, or in, a Euclidean space. It's flat. But that's not what you will find when you measure the distance between New York and Paris. You will need to use a different metric for that, since the surface is curved.

What is the metric for a spherical surface? In other words, given coordinate positions of two points A and B on the surface of a sphere, what is the formula for calculating the distance between them?

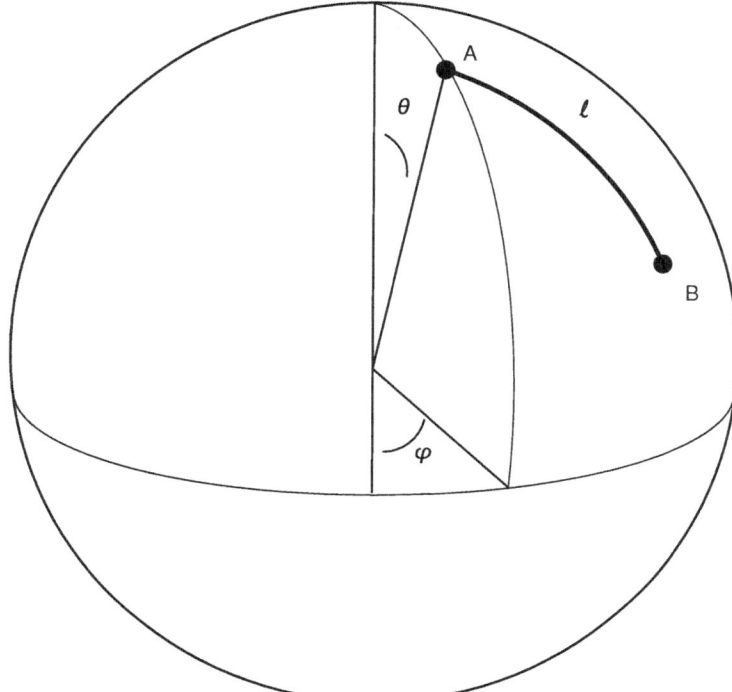

Figure 11.5. Two points A and B on a curved surface, and the distance between them, *l*. The surface is two-dimensional, so two coordinates are required to locate each point; φ is the longitude, and θ is the angular distance from the north pole, the co-latitude. The radius of the sphere is *r*. All points on the surface have the same value of *r*, so *r* is not an informative coordinate.

Start with a specific example in which point A is at the north pole and point B is on the equator. This is similar to an example we did in Chapter 10. The distance from A to B, the length of the interval between A and B, is ¼ of the way around the sphere. It's ¼ of the circumference, so it's $\frac{1}{4} \times 2\pi r$, where *r* is the radius of the sphere. In this case, $l = \pi r/2$. The interval will depend on the radius of the sphere.

To get the more general formula for the metric, for any two points anywhere on the sphere, we need to choose a coordinate system to describe exactly where the points are. We got by with the vague at-the-north-pole and on-the-equator for the specific case, but that won't work in general. Since all points on the surface of the sphere are at the same distance *r* from the center, there is no distinguishing their position by their value of *r*. So *r* won't be a useful coordinate. The two-dimensional surface needs two coordinates to locate a point. Latitude and longitude will do. We'll use φ to represent the longitude and θ to represent what's called the co-latitude, the angular position as measured from the north pole. This is shown in Figure 11.5.

In this coordinate system, the two points have a longitudinal angular separation of $\Delta\theta$, and a co-latitude angular separation of $\Delta\varphi$. On a sphere with radius r, the distance between A and B is given by the following:

$$\Delta l^2 = r^2(\Delta\theta)^2 + r^2\sin^2\theta(\Delta\varphi)^2 \tag{11.4}$$

It's not important (for us) how this formula for the metric is derived, or even how to use it. Only two things about the math are important, and they are related. One is the appearance of θ in the coefficient that multiplies $(\Delta\varphi^2)$. The other is the change from labeling the squared value of the length as l^2 to now Δl^2.

Never mind the r^2. That value remains constant and is the same everywhere on the sphere. But the value of θ is different at different points on the sphere. It's a variable. This is the sign that the surface is curved, intrinsically curved. If the coefficients in the metric are constants, the surface is flat. If any of them are variables, the surface is curved. Since θ changes from point to point, the measure of length changes from point to point. This is the stretching that is characteristic of curvature.

So, what value of θ should we use in the formula, the value at point A or the value at point B or the value at a point somewhere in between? The answer is all of the above. At each point along the geodesic from A to B, use the formula to calculate a tiny increment of distance at that point using that value of θ, and then add up all these incremental distances to get the total, the interval from A to B. That's why there is the Δ in front of the Δl^2. Add up the variable changes in length, the Δl at each point along the path, to get the total distance. To do this properly, the increments would have to be infinitesimally small, and the summation would have to be a continuous integration. This requires calculus, but only to actually do the calculation.

No calculus is needed to understand the physics expressed by the formula. All we need to note is that on this curved surface, the sphere, the metric has a coefficient, a multiplier of one of the coordinate lengths, that is itself a variable. In this case it's the θ in the $r^2\sin^2\theta$ that multiplies the $(\Delta\varphi^2)$. This is generally the case. A variable in the metric is the sign of curvature. It means geodesics will bend and the resulting space will not be the flat geometry of two- or three-dimensional Euclidean space or four-dimensional Minkowski spacetime. So, since gravity is associated with curved geodesics, gravity will be expressed as variable coefficients in the metric.

Since the important information about curvature, and hence gravity, is in the coefficients of the metric, a handy way to write the metric is to give just those coefficients. For example, on the flat two-dimensional surface with perpendicular x, y axes, as in Figures 11.2 and 11.3, the metric is as follows:

$$l^2 = (\Delta x)^2 + (\Delta y)^2 \tag{11.5}$$

The coefficients are simply 1 and 1, that is $1 \times (\Delta x)^2$ and $1 \times (\Delta y)^2$. So the metric can be written as (1,1). The three-dimensional flat surface has a metric (1,1,1). The two-dimensional spherical surface has a metric $(r^2, r^2\sin^2\theta)$. And so on. There is no new math or physics in this step. It is simply an abbreviation to highlight the important features of the metric.

Gravity will be represented by a metric with variable coefficients. But it can't be a metric in two- or three-dimensional space. It will have to include the time dimension as well. This is from Chapter 9, where it became clear that the only way to include the fundamental fact that the speed of light is invariant is to do the physics in four-dimensional spacetime. This is necessary in order to avoid the internal failure of Newtonian gravity, the instantaneous action at a distance. So we need to figure out the appropriate metric for four-dimensional spacetime.

In two-dimensions, a flat surface has flat geometry, that is, Euclidean geometry and the metric is the Pythagorean theorem. In four-dimensions, the flat geometry is Minkowskian geometry. We need the metric, the spacetime analog of the Pythagorean theorem. In the same way that we didn't derive the Pythagorean theorem (leaving that for Pythagoras), we won't derive the Minkowski metric either. We'll just write it down.

$$l^2 = (\Delta x)^2 + (\Delta y)^2 + (\Delta z)^2 - c^2(\Delta t)^2 \qquad (11.6)$$

In abbreviated form, the Minkowski metric is $(1,1,1,-c^2)$, where c is the speed of light.

Again, neither the derivation nor the application of the formula is important for the purpose of understanding gravity and geometry. And again, the important thing is to note whether any of the coefficients in the metric are variables. They are not. They are all constants, including the speed of light. So, Minkowski spacetime is flat. Geodesics are straight. There is no stretching. There is no gravity.

Just as the Pythagorean interval between two points A and B on a flat surface does not change when you change from one coordinate system (Figure 11.2) to another (Figure 11.3), the spacetime interval between two events A and B does not change when you change from one coordinate system to another. The spacetime interval is invariant. And since geometry is determined by distances, in this case intervals between events, the geometry is invariant. The interval between A and B on a flat surface is the straight-ruler distance. On a sphere the interval is the taut-string distance, as a crow flies, the path of an airplane. In spacetime, the interval is the path of light. Light follows the geodesic, and this is expressed by the metric.

Now add gravity to the four-dimensional spacetime metric. That is, now allow for one or more of the coefficients in the metric to be a variable, a variable that is influenced by the massive objects in the area, the Sun, say. In this way, the massive object will stretch the spacetime, curve the geodesics, and, by the Principle of Equivalence, cause the gravity.

Gravity, we know from Chapter 10, affects the passage of time. This is the gravitational red shift, the physical fact that clocks, and all periodic phenomena, tick more slowly where the gravitational field is stronger. For an object like the Earth or the Sun, the gravitational field is stronger the closer you are to the source, and it gets weaker as you get farther away. In the Newtonian theory, the force gets weaker by a factor of $1/r^2$ at a distance r from the source. Putting these pieces together shows that the measure of time, that is, the coefficient for the Δt component in the spacetime metric will have to be a variable that changes according to the distance from the source of the gravitational field. The gravitational red shift of time is (metaphorically) the stretching of the time component in the metric. Gravity, in other words and again metaphorically, curves time. Gravity, in other words but this time literally, makes the time coefficient in the spacetime metric a variable.

The particular form of the variable, the function of the variable, in the time component of the metric, will depend on the details of the particular source of gravity, its shape and mass and motion. This is no different from Newtonian gravity, in the sense that the gravitational field depends on the details of the source. In the case of the general theory of relativity it is a metric field rather than a force field, so it is the details of the metric that are affected. The Newtonian theory of gravity, inspired as it was by planetary astronomy, naturally developed around a massive source that is a sphere, like the Sun or a planet. The question was about the gravitational field at some distance r from the source. We will do exactly the same thing for the general theory of relativity, the metric-field theory we're building.

If you prefer avoiding the mathematical details in putting together the metric, you can safely skip to the boxed equation below, Equation (11.9). The importance of each of the working features of that equation will be explained after the box.

Starting with just the metric coefficient of the time coordinate, the multiplier of the Δt, the gravitational red shift is accurately described by the factor

$$(1 - 2GM/c^2 r) \tag{11.7}$$

In other words, gravity expressed as a spacetime metric that includes the gravitational red shift must be

$$\Delta l^2 = (\Delta x)^2 + (\Delta y)^2 + (\Delta z)^2 - (1 - 2GM/c^2 r)c^2(\Delta t)^2 \tag{11.8}$$

As usual, we don't need to know how this formula was derived or how to use it to calculate exact results. What matters is why the complicated coefficient is in the metric and what it means. It means that the time dimension is stretched by a factor that depends on the mass M of the spherical object, the Sun or the planet, and the distance r away from that object. Because r is a variable, the red shift depends on how far away from the source you are. And since this coefficient in the metric is a variable, the spacetime is curved. Geodesics are not straight, and the cause is gravity.

But we're not done. The special theory of relativity requires that the speed of light be invariant, that is, the same in all reference systems. This means that if the coordinate measure of time Δt is changed in any way, that is, if it is stretched in the way the red shift does, then the coordinate measure of space must change as well. This is what you get in a world in which the absolute is a speed rather than space and time separately, as in the Newtonian model. Speed is spatial distance divided by time, so space and time must change in compensating ways to keep the speed, the speed of light, invariant. The spatial and temporal parameters must co-vary. In other words, since there is the variable coefficient in front of Δt in the metric, there has to be a variable coefficient in front of the Δx, Δy, and Δz.

Adding a variable coefficient to the spatial component to the metric is the fundamental difference between the general theory of relativity and the Newtonian theory of gravity. The Newtonian theory accommodates the red shift, and in that sense it stretches the measure, the metric, of time. But the Newtonian theory does not accommodate the invariant speed of light. The action at a distance, the "uncommon unintelligibility," violates this. So the Newtonian theory does not include the compensating stretching of space. This will lead to different predictions from the two theories. The differences will be small but measurable.

When you do the math, the spatial coefficient in the metric turns out to be $1 + 2GM/c^2r$. It's the same for all three spatial coordinates, Δx, Δy, and Δz, since we are working on the particular case of a spherically symmetric source like the Sun or a planet. Put things all together and we get the full metric.

$$\Delta l^2 = (1 + 2GM/c^2r)[(\Delta x)^2 + (\Delta y)^2 + (\Delta z)^2] - (1 - 2GM/c^2r)c^2(\Delta t)^2$$

$$(11.9)$$

This may be a bit overwhelming in its mathematical complication, so we should call attention to the important features of the equation to clarify why they are there and what they mean.

Most importantly, note the occurrence of the variable r in the metric. That shows unequivocally that the geometry of the spacetime is curved. A variable in the metric means curvature, just like on the two-dimensional sphere. And it's intrinsic curvature, determined within the space itself.

Note also that the curvature is in both the space and time dimensions. This is required of any relativistic theory, that is, any theory that is consistent with the fact that the speed of light is absolute.

And third, the reason there must be the variable r in the time component of the metric is simply the fact of gravitational red shift. And that, we saw in the previous chapter, is fully expected in light of the Principle of Equivalence.

To summarize, we got to this metric theory of gravity, the boxed equation, by starting with the Principle of Equivalence, which we first encountered with Galileo in Chapter 6, and folding in the fact that the speed of light is invariant. The fact about light is itself a consequence of a fundamental principle, the Principle of Relativity. That, too, we saw in Chapter 6. Galileo, again.

It's important to remember that the mathematical formulation of the theory we have at this point, the boxed equation, is for a particular case. We specialized to finding the metric expression of gravity as caused by a spherical, unchanging mass M. The field, the metric field, changed at different distances from the source, different values of r. If the source was more complicated, if it was not spherically symmetric, or if was rotating or otherwise moving, the metric would be more complicated than what we have in the box. In particular, if the source is moving, that is, if things are changing over time, the time variable will appear in coefficients of the metric. The metric will be a function of not only space, that is, r, but also time. The math gets complicated quickly. There can also be cross-terms, similar to the $\Delta x \Delta y$ term that showed up when we used non-perpendicular axes in the coordinate system on the two-dimensional surface. The metric is not just the four coefficients for Δx, Δy, Δz, and Δt, but potentially 16 coefficients arrayed in a 4×4 matrix to cover all possible cross-terms. The math gets complicated quickly. The mathematical term for the matrix is a tensor. In this case it's called the metric tensor. It holds all the information about the curvature of spacetime. And, by the Principle of Equivalence, we understand how massive objects cause that curvature. The metric tensor holds all the information about gravity. That's the general theory of relativity. It's a metric field-theory of gravity.

But now, how does the curvature of spacetime affect the motion of things? Why, in other words, does a dropped object fall? With Newton the answer was pretty easy, as long as you ignore the mysterious action at a distance. With Newton, a dropped object falls because a massive object like the Earth creates a force, $F = GMm/r^2$ and the dropped object falls with acceleration in accordance with the law $F = ma$. But gravity is not a force in the general theory of relativity. It's the curvature of spacetime. How does this cause the dropped object to fall, or the planets to orbit the Sun?

Let's talk about a specific dropped object, an apple. It used to be a simple question to ask if the apple is moving or not, but since the Principle of Relativity showed up in our analysis we realize that the question only makes sense when a particular reference frame is specified. So, work in the reference frame in which the apple is not moving. Do this, for example, in a reference frame out in space, in a box or room far from any massive objects, with the apple just floating in the middle of the box. If there are no forces on the apple, it will stay still and not start moving. No force, no acceleration. An object subject to no force is called a free particle.

The apple that is stationary and floating in the box is not moving in *space*, but it is moving in *spacetime*. Its trajectory through spacetime is a line, its worldline. In the reference system of the box, the apple's worldline through spacetime is straight and, as we saw in Chapter 9, vertical. And, of course, in some other reference system, one that is moving with respect to the box, the apple is moving both in spacetime and in space. In that reference frame the worldline will still be straight but no longer vertical. It will be tipped at an angle determined by the relative speed of the reference frame.

The worldline of a free particle is a straight line. A force would make the apple accelerate and that would curve its worldline. This is true at least in an inertial reference frame. And a straight line in an inertial reference frame is a geodesic. Remember, our box is out in empty space where there is no gravity and hence no curving the metric. The geodesics are simply the straight lines. In other words, free particles, all free particles, follow geodesics through spacetime. It would take a force to steer the particle, the apple, off the geodesic.

What happens if the box itself is accelerating, that is, if the reference frame is non-inertial? If rockets on the bottom start to accelerate the box upward, the apple will spontaneously accelerate to the floor. Only in an inertial reference frame will the free particle remain still.

Back in the inertial reference frame, in the box with the apple stationary and floating with no support, the apple is following a geodesic through spacetime. But this time, the box itself is in free-fall toward the Earth. You're not to blame for not noticing this change of circumstances, since, by the Principle of Equivalence, there is absolutely no physical difference between a reference frame where there is no gravity and one that is in free-fall in a gravitational field, at least no difference on the small scale of the box. So the stationary floating apple has no forces acting on it, just as it didn't in the first circumstance, and it stays still in this reference system. The falling box is an inertial reference system. Inertial reference systems are determined by massive objects, like the Earth, in that free-fall is inertial motion.

Now simply remove the box. There is no force on the apple, just as there was no force on it when it was in the box. And it is in free-fall toward the Earth. And it still follows a geodesic through spacetime, as it did while floating in the box. The free particle follows the geodesic through spacetime, and the geodesic is curved by the massive object, in this case the Earth.

So, why does an apple fall to the ground? It falls because there is no force acting on it, and, like all free bodies, it follows a geodesic path through spacetime, and the geodesic is determined by the nearby mass, the Earth.

An apple sitting on a table, that is, not falling, is moving through spacetime, just as the falling apple is moving through spacetime. But on the table the apple is subject to a force, just one, namely the upward force of the table itself. The force pushes the object off the geodesic trajectory through spacetime. Unlike the falling apple, the free body, the tabled apple does not follow a geodesic. It is at rest in the reference frame of the room, but that reference frame is not inertial. Only free-fall reference frames are inertial.

Notice how the general theory of relativity distinguishes between free particles and those that are subject to a force. Free particles, those left alone, so to speak, follow the geodesics, the straightest lines of the spacetime. Forced particles deviate in some way. This distinction is reminiscent of the Aristotelian distinction between natural and violent motion, described in Chapter 4. Natural motion, in the Aristotelian theory, results when an object is left alone to seek out its natural place in space. The apple, being mostly water and earth, moves beneath the air. It falls to the ground. Violent motion, motion from some outside interference like a lifting hand or a supporting table, keeps the object from its natural place. The apple is forced to stay above the air. In the general theory of relativity, natural motion results when an object is left alone to follow the geodesic in spacetime. The motion in this case doesn't depend on the composition of the object, water or earth or whatever. That's in the Principle of Equivalence.

Geodesics clearly play a central role in the general theory of relativity and its account of gravity. We started the description of geodesics with idea that light follows geodesics, the straightest possible path through spacetime, the taut-string, as-a-crow-flies straight when the spacetime is curved. But now we have free bodies, apples for heaven's sake, following geodesics. How is that possible? Light goes really fast. Apples go comparatively much slower. If I shine a light and release an apple from the same place at the same time, they are certainly not going to follow the same trajectory. No, they're not. But just as through any one point in space there are lots of straight lines, through any one event in spacetime there are lots of geodesics. Which one is followed, by the light or the apple, depends on the initial conditions, the initial speed of the object or the light. The dropped apple starts with speed zero, and that determines which of the geodesics it will follow. Light starts with speed c, and that puts it on a different geodesic. In both cases, though, light and free particle, the trajectory is a geodesic in spacetime, a geodesic shaped by gravity, that is, by the local distribution of mass. In our case, that mass distribution is the Earth.

Planets, in their orbits around the Sun, follow geodesics, too. So does the Moon and so do satellites and the International Space Station. They are all free particles. Their initial velocity determines which geodesic they will follow, just as the initial angle of striking a cue ball determines which of the many straight lines it will take from the point of contact. Light is fast; it follows the one geodesic that matches that speed. Using the descriptions of spacetime intervals from Chapter 9, this is the lightlike geodesic. Planets and apples are slow. They follow timelike geodesics.

The term timelike makes sense. Since planets and apples are slow, they don't change position in space very quickly. Their Δx, Δy, and Δz are minimal during an interval of time Δt. Consequently, the spatial component of the metric, the stretching of space, has little effect. This means that the temporal component of the metric, the stretching of time, has the dominating role in shaping the geodesics they follow. This also means that the Newtonian theory of gravity, since it includes the phenomenon of gravitational red shift and consequently accommodates the stretching of time, is a good approximation for the gravitational behavior of slow objects like planets and apples. It's no surprise that the Newtonian theory did so well, and continues to do so well, with planetary astronomy, trajectories of projectiles, and rocket science.

Light is fast. That means its trajectory through spacetime, its worldline, covers a lot of space in a little time. It has a significant Δx, Δy, Δz in an interval of Δt. Consequently, the spatial component of the metric has a significant effect. And since it's on the spatial coefficient that the general theory of relativity and Newtonian gravity differ, it's in the behavior of light that the two theories will make significantly different predictions. That's where we should look to test the new theory. Both theories predict that light will bend in a gravitational field, since the variable coefficient of the time component of the metric, the red shift, bends the geodesic. But they differ in how much the light will bend, since the general theory of relativity adds the variable coefficient of the spatial component. Measuring the bending of light, the amount of bending, will be the test.

Planets are slow, but if the gravitational field is very strong, the spatial component of the metric will be significant. The metric depends on M, the mass of the central body like the Sun or the Earth, and $1/r$, the inverse of the distance to the central body. Close to a very massive body, the stretching of both time and space is considerable. Close to the Sun, a planet will be affected by both time and space curvature. Mercury, in other words, among all the planets, is the one that will show a deviation from the Newtonian theory, since that theory has no spatial curvature. The orbit of Mercury will be another test of the new theory.

This brings us up to date. The general theory of relativity is the current account of the fundamental nature of gravity. We will test the theory in the next chapter. It's the theory you learn in graduate school as a physicist. This one is not merely good enough for engineering and rocket science, it's considered to be true in all detail. Now we know. What goes up . . . must follow the local timelike spacetime geodesic that is curved by the nearby mass.

12

Testing the General Theory of Relativity

Aristotle is often dismissed by modern scientists and science writers as being not only wrong about gravity, and a lot of other things, but much worse, not even scientific in his efforts. There is no denying that he was a giant, but maybe more in the sense of being a big bad giant in the way, rather than a hero with shoulders on which to stand and see further toward the horizon of truth. Bernard Schutz, who begins his fantastic book on gravity with Galileo, mentions Aristotle in the preface. "Aristotle's view that objects fall to the ground because they are seeking their natural place was a big obstacle when Galileo began to formulate laws of motion and gravity mathematically" (Schutz, 2003, p. xxv). In fact though, Galileo found much to admire in Aristotle. He retained the perfect circles as the natural celestial motion, and he implicitly kept the vertical straight line as a natural motion on the Earth. The accomplishment was to allow an object to have *both* the circular, that is, horizontal, and vertical natural motions.

There were matters of method that Galileo found insufficient, matters in which Aristotle, or at least subsequent Aristotelians, were in the way. Consider the basic Aristotelian method for understanding nature. There are, of course, variations on this theme, but it starts with fundamental generalizations about how things are and how they change. Nothing has more than one natural motion, and so on. From the basic principles, Aristotle moved by the logic of deduction on to more particular claims about the nature of specific things and phenomena. This is arm-chair science, with conspicuous disregard for evidence. It is also exactly the method used in what Einstein called a theory of principle, like the theory of relativity.

Given this apparent similarity in method between Aristotle and Einstein, it should be interesting to compare the two scientists in some detail. The historical narrative of the science of gravity, the description of sequential changes, makes it easy to contrast one major theory to the next, Aristotle to Newton and then Newton to Einstein. It requires an extra effort to view Aristotle and Einstein side-by-side, but the effort pays off.

The methodological complaint against Aristotle is not so much about what he did; it's about what he didn't do. The abstract and analytic derivation of a theory is compatible with proper scientific method, as long as the derivation is followed up with empirical testing. Aristotle, and more importantly those who subsequently adopted and advocated his ideas, failed to follow up. There is no requirement that each individual scientist empirically tests his or her own theory. No one is obligated to do it all on their own. But scientific endorsement of a theory demands that *someone* in the group does the testing. The requirement extends to the theory of relativity as well. It's not that Einstein himself had to roll up his sleeves and test the theory. But the theory needed to be tested. It required new evidence, that is, data that had not been used to motivate the theorizing. Even a theory of principle must confront the risky outcome of prediction. This is what we need to look for to distinguish the method of relativity from the method of the Aristotelians.

First, hold all three of the major theories of gravity together up to the light and see what they say and how they put it. First ask of each theory, does it make sense. We'll do this both for our own sake, to make sure we understand the big ideas, and for the sake of science. It should be a basic requirement of a theory that it makes sense, nonsense being scientifically inadmissible. Well, maybe not, given Mach's description of the Newtonian theory of gravitation as first "uncommon unintelligibility" and then "*common* unintelligibility" (Mach's emphasis). Perhaps he was suggesting that Newton was no scientist, at least not in his work on gravity. Or is it that intelligibility is unnecessary for acceptable science? What is the role of making sense in the context of justification of a scientific theory?

Making sense, intelligibility, can be a personal accomplishment or failure. What makes sense to me might not make sense to you. Making sense is, in this way, a psychological and subjective property of a theory and theorist. It is also relative to a historical and intellectual context, and to the particular set of beliefs at the time. It's relative, in other words, to a scientific paradigm. This suggests that intelligibility may be an unworthy criterion for evaluating whether a theory is likely to be true. It may have no legitimate place in the method. But there is a less flexible measure of intelligibility that is not under any idiosyncratic influence. A theory has to be internally consistent or it is nonsense. This is a matter of logic, and it can be assessed and enforced in an objective, impersonal way. If there are logical contradictions in a description of nature, the description cannot be true, at least not entirely true. Nature does not accommodate contradiction so neither can science.

Intelligibility as internal consistency is a requirement in any scientific theory, but it's a pretty weak requirement. Fairytales are internally consistent. So we want a stronger measure of "makes sense" in science to count as not just a necessity but a virtue, a contributing reason to think the theory is true. The stronger standard is in terms of implications between claims. Call this **coherence**. There is an important logical difference between saying that two statements can both be true, that's con-sistency, and saying that the truth of one implies the other. A system of descriptive claims, a theory, makes good sense if it is not only consistent but interconnected such that this is a consequence of that and implies the other, and so on. Rather than just a list of facts, a coherent theory is a network of facts such that we know not just the building blocks of nature but we understand the architecture as well. This is the greater achievement, and under the assumption that nature is in fact systematic and organized, a better indication that the theory is getting to the truth.

The feature of a scientific theory we're calling coherence needs work, both to make it more precise and measurable and to secure its status as a theoretical virtue, at least some small symptom that the theory is likely to be true. But at this point, the broad notion of consistency and connection among the components of a theory will help frame the comparison of Aristotle, Newton, and Einstein.

The core component of Aristotelian gravity is the distinction between natural and violent motion. Natural motion is guided by a point in space, the center of the cosmos, and by geometry. Natural motion of a terrestrial object is straight, that's the geometry, and either toward or away from the center of things. Natural motion of a celestial object is circular, geometry again, centered on the same point. In these terms, there are two versions of gravity, terrestrial and celestial, but the guiding principle is shared. It is in the nature of a stone to fall; left alone, without violent interference, it follows its natural motion straight toward the center of the universe. A planet orbits as its nature is the circular path. There is no violence to disrupt celestial motion. As a system of ideas, there is no internal contradiction here, no nonsense.

Note the similarity between Aristotle and Einstein in the basic concepts deployed in describing gravity. General relativity depends on a distinction between a free particle and an object with a force applied. A free particle is guided by a line in spacetime, a geodesic, and by geometry. A dropped stone or an orbiting planet is free to follow the local geodesic, and if it weren't for the conspicuous comparison to the Aristotelian account, we might add that that's its nature. Unlike the Aristotelian account, the interaction is local rather than teleological. The constraints on how to move are right there at the location of the particle, not at some distant goal. And unlike the Aristotelian system, the geometric features are dynamic in that they change, both in time and place, and they are affected by the objects in the cosmos. There are these important differences, but there are also those core concepts that are shared.

Now look closely at the methods, Aristotle's and Einstein's. Aristotle, we have noted, started with basic principles like natural motion and perfection in the heav-ens. But the principles were not invented on a whim or by blind speculation or some philosophical preconception. There was some evidence. Things fall down, and usually heavy things hit the ground a little sooner than light things. To say this is their natural motion is to generalize on empirical data. And at the time, astronomical observations revealed no changes in the sky, year in and year out. Celestial existence and motion looked to be eternal, unimpeded, and flawlessly repetitive. Thus, the expectation of perfection.

This is old evidence, in the sense that it had been observed and recorded before coming up with the theory. The theory explains phenomena already seen, and this is its empirical base. What's missing is new evidence, the prediction of data yet unseen or at least unmeasured.

General relativity is likewise supported by old evidence. It explains the usual gravitational phenomena like free-fall and orbits. And it explains the previously measured precession of the perihelion of the orbit of Mercury. But general relativity also confronts new evidence, and this is the methodological difference. We need to ask just how big a difference this makes, between explanation and prediction, not just in the case of gravity but in scientific method generally.

The Newtonian theory was empirically rich in both existing data, quantitative and precise with Galileo's laws of free-fall and Kepler's laws of planetary orbit, and predicted new data of both terrestrial and celestial trajectories. Newton was an empiricist, with primary allegiance to the evidence and unwilling to theorize much beyond. Aristotle and Einstein shared the sensibility of theorists, prioritizing the conceptual and logical virtues of their descriptions of gravity over the details of the connection to evidence. Testing the theory seemed unnecessary for Aristotle and simply a formality for Einstein. When the mathematics revealed that his general theory of relativity matched the precession data for Mercury, Einstein felt full triumph of his theory: "For a few days I was beside myself with joyous excitement." (Quoted in Schutz, 2003, p. 238.) It was not just solving the old empirical problem, it was the theoretical beauty and conceptual inevitability of the theory, the theory of principle, that had him convinced.

But science is not about personal sensibilities and convictions, even Einstein's. It's about institutional standards and process. Science the institution continued beyond the explanation of Mercury's orbit and the theoretical coherence of the general theory of relativity and asked for new evidence. It went on to test the theory. It's not an easy theory to test, since its distinctive predictions require astronomically strong gravitational fields not found on Earth. So it was back to astronomy to test a theory of gravity.

The gravitational red shift of electromagnetic radiation is in fact measurably large, even in the terrestrial setting. But every theory of gravity, including the Newtonian, predicts not only the fact but the same amount of red shift. The phenomenon has been measured, but it is no distinction for the general theory of relativity. The most famous test of the theory, and the first after its introduction in 1916, was the measurement of gravitational lensing. Again, any theory of gravity that abides by the Principle of Equivalence includes the phenomenon of gravity bending light, but different theories differ in the amount of bending. So, it's the amount of bending you need to predict and measure to test one theory in opposition to another.

The classic test of the theory uses the Sun, not as the source of the light to be bent but as the massive source of the gravitational field. A ray of light from a distant star will go straight unless it passes very close to the Sun where the spacetime curvature will deflect the ray inward just a bit. Detecting this requires comparing the visual position of the star when the Sun is nowhere nearby to the position when the Sun is practically in the way. The latter measurement requires the event of a solar eclipse, with the Moon blocking the light of the Sun so that nearby stars are visible. Einstein and the scientific community had to wait, and then they had to go to a place on the Earth where the eclipse would be total.

The general theory of relativity makes a precise prediction for the degree of deflection, given the mass of the Sun and the distance at which particular starlight will pass. Einstein had a first draft of the theory and a numerical prediction of deflection in 1911. The first opportunity to test the prediction, the first accessible solar eclipse, was in 1914. A group led by the German astronomer Erwin Freundlich was sent to the Crimean peninsula where the eclipse would be visible. It is challenging to control experiments and observations outside the sheltered conditions of a laboratory. The Freundlich expedition was prepared for the vagaries of nature, but perhaps not nations. They were disrupted literally by an act of war, World War I. The scientists were detained and their astronomical instruments, mistaken for weapons, were confiscated. No data were gathered.

With hindsight we can say that this delay in testing general relativity was a lucky break for Einstein and the theory. No lives were lost; the scientists were released unharmed and reunited with their equipment, after the eclipse. But more importantly, the early version of the theory, and Einstein's calculated prediction of starlight deflection, expected a value of 0.85″. After 1914, with some revision of calculation, Einstein changed the prediction to twice that, 1.75″. The Crimea expedition was testing a prediction that was itself off by a half. If they had succeeded in accurate measurements, the results would have disagreed with the theory. What conclusion would the scientific community have reached? What would Einstein have done? He is famous for a bit of theoretician's hubris after the later, supportive test. When asked in 1919 what he would have thought if the test results conflicted with his theoretical predictions, he replied, "Then I would feel sorry for the good Lord. The theory is correct anyway." (Quoted in Ohanian, 2008, p. 4.) Easy to say when the scientific community, the reviewing peers, are celebrating the agreement of prediction and evidence. But even in 1914, before there were test results, Einstein was confident: "Now I am fully satisfied, and I no longer doubt the correctness of the whole system, whether the observation of the eclipse succeeds or not." (Quoted in Clark, 1972, p. 222.)

Speculation on personal, hypothetical history is not useful except to draw attention to the uncertainty and resulting room for discretion in empirical testing. A single failed prediction, as would have happened had the 1914 measurements been done and done accurately, does not disprove a hypothesis. It shows something is wrong somewhere, in the basic theory, or in the specifics of derivation, or in the conduct of the observation, or in the interpretation of the data. Testing is neither immaculate nor decisive. This is evident as well in the test that did match the prediction, the famous eclipse measurement of 1919.

By 1916, Einstein had revised the derivation of gravitational lensing by the Sun, and the prediction was 1.75″. The key is to include the curvature of both the time component of the metric, as he had done initially, and the space component, which he had neglected. A solar eclipse in March 1919 provided the astronomical circumstances to test the new prediction. This time the team was led by a British astronomer, Arthur Eddington. There is reason to think that Eddington, a pacifist, was very much hoping the evidence would support the hypothesis. It would be a reassuring example of a German theory proven by a British effort, demonstrating the value and promise of cooperation between previous rivals.

Measuring the gravitational effect of the Sun on the trajectory of starlight is easy in principle. It's a matter of comparing photographs of a particular group of stars with and without the Sun in the way. But getting the photographs, particularly the ones during the eclipse, and then making careful sense of the comparison is fraught with difficulties. The eclipse will not come to the observatory, with its high-quality, stable telescopes. The observations must be made in the field, with portable instruments and makeshift conditions. Small telescopes require long exposure times, and that entails steady tracking of the telescope to compensate for the movement of the Earth. The eclipse exposures are made during the day, when the telescopes are warm, but suddenly cooling during the brief event, while the comparison photographs are made at night, with evenly cooled telescopes. And, of course, there is the weather. Normally, an astronomer can wait out a cloudy night and make the measurements on the next, but the eclipse won't wait. A cloudy day means a missed opportunity and no data.

The Eddington expedition suffered all of these challenges. Two groups were deployed, hoping to spread the risks. One, with Eddington himself, set up on Principe Island off the west coast of Africa. Another was sent to Sobral in Brazil. It was raining at Principe on the morning of the eclipse. Only after the event had begun did the clouds part and allow for photography. Sixteen pictures were taken, but only two of these were clear enough to use. Five stars were identified and their positions measured. Sobral had better luck with the weather. Two sets of photographs were taken, with two different devices, with differing quality. A four-inch telescope produced eight photographs, with seven identifiable stars, while a shakier astrograph, the same sort of instrument used by Eddington, gave 18 photographs.

It's not just the fuzziness of the photos that made these data difficult to deal with. The two plates from Principe showed an averaged deflection of 1.6″. This is in rough agreement with the predicted 1.75″, but it is just the average of two values. The two sets of data from Brazil were equivocal. The eight photographs from the telescope averaged 1.98″, but the eighteen from the astrograph gave 0.86″, in agreement with Einstein's earlier prediction and with the prediction from Newtonian theory. Despite the dodginess in the data, Eddington, and the international press, saw confirmation of the general theory of relativity. This was the event that made Einstein famous and his name a synonym for genius. Newton was wrong; Einstein was right.

The enthusiasm among non-scientists is understandable and excusable. But it not only ignores the cloudy evidence on a cloudy day, it is a clear fallacy of affirming the consequent. The theory is not proven. Despite this, even many scientists overstate the success for the theory of relativity. Clifford Will, who is careful to point out the troubles in both gathering and interpreting the eclipse data, concludes that this was nonetheless "a victory for relativity" (Will, 1986, p. 78). And James Cushing summarizes the test simply as "stunning, positive results of both expeditions" (Cushing, 1998, p. 257).

High-profile, one-chance-only results are vulnerable to this kind of over-reaction that commits the fallacy of affirming the consequent. When there is just one machine that can run the experiment, like the Large Hadron Collider used to look for the Higgs boson, or just one special event to provide the circumstances to check a prediction, the standard of repeatability of testing tends to be overlooked. The case of gravitational lensing, luckily, allows for some repetition and follow-up, as there were solar eclipses after 1919. But subsequent eclipse data were not much better, and it wasn't until 1969, using radio rather than optical telescopes, that the results were significantly more reliable and consistently supportive of the general theory of relativity.

The data improved, but the logic stayed the same. Whether it's testing by predicting new observations or explaining old observations, the logical link between theory and observation is always that the theory entails the observation. It's never the other way around. The general theory of relativity entails a 1.75″ deflection of starlight, and it entails the extra 38″ precession of Mercury. Einstein seemed more impressed by his success with the old evidence and dismissive of the new. But most descriptions of scientific method emphasize prediction, and some are even dismissive of explanation. Intuitions differ on the relative strengths of old and new evidence. To see if one is in fact more important to the method, we need to find the reasoning inside the intuition.

Why favor new evidence over old and claim that a successful prediction is stronger evidence in favor of a theory than what is sometimes called retrodiction? The answer is that it seems too easy to come up with an explanation that accommodates the existing evidence. It's like knowing the answers first, or seeing the points on the graph and simply drawing the line that hits them all. Any number of lines can be drawn through those points, and there's no limit besides imagination to the explanations of recorded phenomena. It is better, because it is more difficult and much riskier, to make an informed forecast of the next point on the graph. The prediction could be wrong in a way that retrodiction cannot, and this is what makes a theory falsifiable. That's the case for new evidence over old.

But once that new data point is on the graph, there will again be an unlimited number of lines that go through it and all the previous points. New evidence becomes old evidence, and it then suffers the same underdetermination that makes it impossible to confirm a theory. And, had the theory failed in its prediction and the line missed the new point, that still does not, as a matter of practice and a matter of logic, show the theory is false.

There is no logical advantage to new evidence over old, and there is a significant practical disadvantage. New evidence, observations that are made with the theory in mind, will be interpreted under the influence of that theory. This is pretty clearly what happened to Eddington. That may be an unusually egregious case of allowing the defendant to pick the jury and edit the evidence, but the process of science is often set up to allow, if not encourage, this kind of reciprocity between theory and evidence. Scientific evidence is never crystal clear, even without the troubles with weather, war, and wobbly equipment. There are always good data and bad, and decisions have to be made. If the outcome is as predicted, or plausibly interpreted to be what was predicted, the experiment is over. If the outcome is contrary to prediction, there is a tendency to keep going. This is surely what happens in teaching labs, an important component in training scientists how to do science. There is a targeted outcome, and you do the lab work until you get it right, that is, get what was expected. The quality of the work is judged, at least in part, by the outcome. Insofar as this is teaching scientific method, it reveals that theoretical expectation makes the empirical work weighted in favor of confirmation. New evidence is knowing the answer, the theoretical answer, before the test.

Old evidence is not so supple. These data could not have been selected or construed to the advantage of a future theory. They are colder, harder facts. And it's not all that easy to find a theory that not only fits but explains the existing evidence. Ptolemy's model of the planets fits the fact that the elongation of Venus, its angular separation from the Sun, never exceeds 45°, but the Copernican model explains it. Ptolemy had enough adjustable parameters that he could fiddle and fine-tune until he had a match. Copernicus had no choice; given the orbital arrangements, the maximum elongation was fixed. Ptolemy was consistent with the old evidence; Copernicus entailed it. Similarly, the general theory of relativity makes the precession of Mercury inevitable, not simply allowable. Furthermore, Einstein clearly did not use the data on Mercury in his derivation of the theory. He did not simply draw the line through the points on the graph. The theory follows from principles, not evidence. So, the answer may have been there at the back of the book, but Einstein didn't look until he was done with his homework.

Old evidence may be less susceptible to manipulation and cherry-picking, but it is not fully immune. It can be edited, re-interpreted, or demoted in importance. Nothing in science is immutable, even a report of what was observed or what to make of it. The anomaly in the orbit of Mercury was just one of several empirical challenges to the Newtonian theory of gravity at the beginning of the twentieth century. As with any measurement, there was room to question the reliability of any of these, if not the numerical results then, as with Uranus, the understanding of the relevant conditions. There was, and still is, some uncertainty about the shape of the Sun and the amount it bulges at its equator. This would affect the precession of Mercury. But once the general theory of relativity precisely entailed, and hence explained, the extra 38″ of precession, uncertainty about the bulge was more or less dismissed. The evidence from Mercury was promoted in importance and, from the new theoretical perspective, recognized to be key. Other phenomena on which Newtonian theory, and the general theory of relativity, were a bit off were seen to be the result of not knowing the details of the objects, comets and the Moon, for example, details that affected the measurements but not the governing theory of gravity. The Mercury evidence was accurate, important, and in agreement with the new theory.

The point of this analysis of old and new evidence is not to be critical in the sense of revealing the dirty laundry of scientific method. The point is to put all components of the method in play and make sure we're aware of how they contribute and how they might deceive. Greater magnification on the idea that theory is based on evidence reveals that there is a reciprocal relation between theory and evidence, and some important fine structure of credibility within each. Theories gain support in some ways that are independent of evidence, as with the conceptual virtues that Einstein and colleagues saw in the principled derivation and resulting coherence of the general theory of relativity. And there is some distinction between old evidence and new, and the inevitable interpretive influence of theory on observation. The logic is the same, but the quality of the evidence may differ.

There is a third category of evidence worth pointing out, a significant subset of new evidence. Call it novel evidence. This would be a prediction of a completely new phenomenon, never imagined until it was suggested by some theory. It's not just a new value of a known property of an observed phenomenon. The novelty is qualitative and the phenomenon is unique to the theory in question. This is not simply the next point on the graph; it's a new graph. And, because the distinction is qualitative, the evidence is less amenable to interpretation. Either this effect happens, or it doesn't. And if it does, and only this theory predicted it would, it would be an extraordinary coincidence and almost impossibly good luck if the theory is false.

The prediction of gravitational lensing, and the deflection during an eclipse in particular, was new evidence for the general theory of relativity, but it was not novel. The phenomenon, if not the amount, had been expected even with the Newtonian theory. But gravitational radiation, or more colloquially, gravitational waves, was new to relativity. It's a uniquely relativistic phenomenon, a novel prediction with which to test the general theory.

Gravitational radiation is a consequence of the very conceptual core of the theory of relativity, the fact that no causal influence can propagate faster than light. A causal connection cannot be instantaneous. If there is a change in the source of gravity, that will affect distant objects, but not instantaneously. There will be a delay as the information and energy associated with the change travel from source to effect. It's a change in the field that moves. It's a wave that propagates out from the source.

The fact that the existence of gravitational radiation is linked to such a fundamental concept is both good and bad. It's good in that testing this prediction, trying to detect the gravitational waves, will be a test of not just a small detail that might be readjusted to accommodate the test results. It will be a test of an essential piece of the theory's coherence, a piece that really can't be removed or reworked without threatening the entire network. The test will be a genuine risk. The downside is that any theory of gravity that is consistent with the special theory of relativity will include gravitational radiation. Being so basic, the phenomenon is not unique to the general theory of relativity. But it does distinguish the theory from Newtonian gravity, and in that sense, looking for gravitational waves will be looking for novel evidence.

Gravitational waves, like electromagnetic waves, are moving patterns in the field. In the general theory of relativity the field is the metric, the variable coefficients in the metric. It's not a force field, because gravity is not a force. Specific changes in the mass causing the curvature of spacetime will result in changes in the curvature, the coefficients of the metric, that propagate out. The changes are perpendicular to the direction of propagation. That is, the gravitational wave is transverse, not longitudinal. In this way it's like an electromagnetic wave or a wave on a string, but unlike a sound wave.

A star that collapses or explodes symmetrically outward will not cause any gravitational waves. The gravitational effect of the star will continue to be indistinguishable from a point source of mass all located at the center. This kind of change in the
source is radial, along any line a wave could propagate. It's a longitudinal change,
but it creates no disturbance in the field. This is why gravitational waves will not
be longitudinal. Only lateral motion of the source will cause a deformation in the
field. The wave must be transverse.

A pair of stars in mutual orbit, a binary star, will cause gravitational waves. This
is analogous to the oscillating electric charge that creates an electromagnetic wave.
The effect in both cases is something different from the static field of attraction,
or possible repulsion in the case of electricity. It's not like the force between two
charges, or the spacetime curvature that draws a stone to the ground and holds a
planet in orbit. Those fields don't change over time, and they require no input of
energy to sustain. But accelerating the source takes energy. That energy goes into
the wave and is radiated away from the accelerating mass or charge.

Gravitational radiation is a time-varying change in the metric. This will have no
effect on a single point particle it encounters, since there is no sense of distance or
measurement at a single point, and that's what we mean by metric. The effect is
a change in the distance between two or more objects, or between two points on
an extended object. This will change the shape of the extended object, distorting
it in the plane transverse to the propagation of the wave. So, if we are looking at
the source, the orbiting binary system, the changes at our receiving end will be
left–right and up–down, but not front-to-back. A solid object will change shape,
periodically bulging sideways then vertically. This is how gravitational radiation
might be detected, by measuring the deformation of a solid mass. But the effect will
be very small. The quantitative prediction from the general theory of relativity for
the case of a hugely massive and briskly orbiting binary source is of a deformation
on the order of 10^{-20}. That is a change of size that is a tiny fraction of the overall
size of the detector. This makes the detection extremely challenging. Attempts have
been made, but with only controversial results.

The better hope for detecting gravitational radiation is to catch it at the source, at
the binary system itself. Again, when the phenomenon is too small to detect in the
lab, you turn to astronomy. If there is no external input of energy to replace what's
radiated away, and, of course there isn't in an isolated binary star, the system will
lose energy. That's what to look for, the loss of energy.

The most famous case of detecting this kind of phenomenon was in 1974 by
Joseph Taylor and Russell Hulse. The key was the discovery of a peculiar pulsar. A pulsar is a neutron star that, as the name suggests, pulses on and off like a
lighthouse. With this one, the frequency of the pulses changed periodically, indicating a Doppler shift towards us, then away, over and over. The neutron star was

in orbit around an unseen companion, likely another neutron star, with an amazingly short orbital period of only eight hours. It was eight hours and decreasing, and the decrease is the evidence of gravitational radiation. The energy lost as radiation would result in a shrinking orbit and hence a smaller period. The binary system was running down as it spent energy making waves. It's a stunningly precise measurement. The period decreases by 2.435×10^{-12} seconds per second. The general theory of relativity predicts a decrease of 2.44×10^{-12}.

The calculation of the predicted value is not at all straightforward. It requires knowing the masses of the two neutron stars, data that have to be teased out of measurements of the precession of the orbit and red shift of the electromagnetic radiation. It's a full workshop of relativistic effects, and the entire theoretical tool kit is used in the calculation. The result is a prediction of a novel phenomenon and a precise agreement with the measured orbital decay. The result for Taylor and Hulse was a Nobel Prize in 1993.

This is, of course, not proof of the general theory of relativity, nor even of the reality of gravitational radiation. It's not observation of the radiation. Better to call it detection, and add that it's very indirect. It's common in science to observe one thing and call it evidence of another. In fact, it's implicit in the basic idea of evidence, as something is always evidence *of*... It points to something else, because it implies something else. Clarifying scientific method is, in part, articulating the logic of that implication. In the case of the binary-pulsar evidence of gravitational radiation, it's a long line of inference. The shifting frequency of the pulsar implies it's in orbit. The decreasing period of the shifting implies the orbit is speeding up and hence – another implication – shrinking. That means the system is losing energy. Gravitational radiation would explain the loss of energy.

There are other explanations. If one or both of the orbiting objects is deformed by tidal effects of the other, this costs energy and the orbital radius would change accordingly. So it would be nothing so exotic as gravitational radiation but as common as the decreasing energy in the Earth–Moon system that goes into moving the oceans up and down. Neutron stars, though, are very rigid and resistant to tidal deformation, so the gravitational radiation may be a better explanation. It may in fact be the best explanation. The phenomenon of radiation fits coherently within the tight network of theory that is general relativity. It's not just consistent with the theory; it's entailed by the theory. Overall, general relativity is principled and well-structured. It has these internal theoretical virtues. And the precision with which the theory predicts the new phenomenon of radiation is a particularly valuable empirical virtue.

With these virtues of the general theory of relativity out in the open, we can start to generalize about scientific method. The magnified image of a method distinguishes between the theoretical and observational components of science. Surely the dichotomy is imperfect, and some things are borderline – we talk of germ theory but a microscope makes most pathogens observable – but the basic separation between what can be observed and what cannot is nonetheless useful. Curved spacetime is not observable, and in that way it is a theoretical phenomenon. This is not to say it's unreal, only that our knowledge of it is not directly empirical. The same status applies to the natural goal of a falling stone, in the Aristotelian theory, or the force of gravity in the Newtonian. Observables are things like the basic kinematics of motion, where things are and when, orbits and trajectories and periods of orbits.

Scientific reasoning is in making connections. It's not just links between theory and observation, but also within a theory. Theory should be consistent with observation or, much better, entail observation. And theoretical claims should be consistent with other theoretical claims or, much better, entail other theoretical claims. Entailment within the theory demonstrates an inevitability of the parts. Once the fundamental principles of nature have been articulated, the rest follows naturally, without having to be added or adjusted by hand. This was a virtue of Newton's universal gravitation, that he started with the approximation of circular orbits to derive the inverse-square force, and elliptical orbits showed up automatically. Special relativity started with the invariance of the speed of light, and the relativity of both spatial and temporal properties followed. The general theory connects the Principle of Equivalence, the identity of gravitational and inertial mass, the metric field, and the inverse-square relation all into an essential network.

So, why think the general theory of relativity is true? Steven Weinberg has a prosaic summary. It is "a tangled web of theory and experiment" (Weinberg, 1992, p. 104). No single aspect of the theory, no one test, no individual virtue can on its own be reason to believe a theory. It's the tangled web. And that means there is no tidy algorithm for deciding when a theory is true. We know the pieces that need to be assembled, but there are no step-by-step assembly instructions and no indication that the assembly is complete.

There is one more piece of the method to consider, one more category of reason to believe a theory is true. It's not testing a theory but *using* a theory. Germ theory for example, we use all the time, washing our hands, sterilizing medical equipment, prescribing antibiotics, and it works really well. It helps us avoid disease. How could it work so effectively if it were not true and if there were not real germs? In the next chapter we'll try this strategy with gravity, using the theory and asking if that adds more reason to think the theory is true.

13

Using the Theory to Explore the Universe

A visit to the Campo de Fiori in Rome puts you at the spot where, in 1600, Giordano Bruno was burned at the stake by the Inquisition. The details of his conviction have been lost but, by all accounts, the heresies for which he was condemned included the claim that the Earth revolves around the Sun, as per Copernicus, and the idea that the universe is infinitely large, commensurate to the powers of God. He also suggested that each star is a Sun like our own, around which orbit planets like the Earth. Whatever his reasoning for this last hypothesis, what are now called **extra-solar planets**, or more simply **exoplanets**, seem a natural continuation of the Copernican revolution. The Earth is not cosmologically special, and neither is the Sun or the Solar system. If there are other planetary systems, there are probably other Earth-like planets that could support life. Not only do we reside in an unremarkable part of the universe, we may not be alone.

The idea of planets in orbit around other stars is at once exotic and familiar. Given what is known about how planetary systems are formed, exoplanets should be quite common. The necessary conditions are not so delicately balanced as to be improbable, and the essential organizing influence is simply gravity. Within the past century, with this theoretical underwriting, there are clear scientific reasons to endorse Bruno's hypothesis. And now, within the past two decades, there are also compelling empirical reasons.

As reported by NASA, there are now almost 2000 confirmed exoplanets, and several thousand more candidates, many of which the agency expects will be confirmed. And the search is ongoing. Weeding out the false positives, that is, the process of confirmation, is an issue of methodology. Very few of the confirmed exoplanets have been directly observed as an image of the planet itself, discernable with a telescope. This became possible in 2013, and it can find planets only if they are very far from the star, that is, very unlike the Earth. Before 2013, and still for the majority of exoplanets, there is only indirect detection, even for the so-called confirmed cases. The unseen planet moves the star, and it's the star's motion that we detect. The connection between the star and the planet is, of course, gravity. Our ability and confidence in using the stellar data as planetary evidence depend on our understanding of gravity.

Even by indirect means it is possible to know not only that there is an exoplanet associated with a particular star but also some of the planet's important properties. Evaluating its ability to support life requires at least approximate values of the planet's size, density, and distance from the star. These data are derivable from the basics of orbit, the mass and the time it takes for a complete revolution. The first exoplanets detected were big and close to the star. They were too close to support life. But now there are confirmed reports of planets the size of Jupiter and the Earth, and recently as small as Mars, in orbits comparable to our Solar system. Again, information on these properties is filtered through a theory of gravity. It will be accurate information, but only if you use the right theory. This raises a concern about corroboration. If you use an inaccurate theory to interpret these data, you will get inaccurate results regarding important things like the mass of a planet. But how could you tell? If all the information on the mass of the planet is interpreted through your theory of gravity, there will be no independent way to see if the interpretation is correct. The role of independence is important in scientific method, and this is the place to see if it can be done.

There are several ways a star might reveal that it holds a planet in orbit, and so there are several detection techniques for exoplanets. One of the most productive is to measure the star's radial velocity, its motion either towards us or away from us. This is in contrast to tangential velocity, the motion of a star across our field of view. Radial velocity can be measured by means of Doppler shift. Elements in the star produce characteristic spectra with well-known emission lines. If the star is moving towards us, these lines are shifted to higher frequencies, a blue shift. If it is moving away, the shift is to lower frequencies, a red shift. The amount of shift is correlated to the speed of approach or recession.

A star with an orbiting companion will wobble. In fact, as we've known from Newton's third law of motion in Chapter 7, the star will orbit the center of mass of the star–companion system, a point known as the barycenter of the system. From a perspective on the same plane as the orbit, the motion of the star will be periodic, sometimes toward the viewer, then away, then toward, and so on. Thus, a measurement of periodic blue shifts and red shifts indicates that the star is orbiting some barycenter. As an example, in a planetary system with the Sun and only the planet Jupiter, the Sun orbits a point that is just a bit outside its own surface. It takes 11 years, the period of Jupiter's orbit, to complete one orbit, and this translates into an orbital speed of 12.5 m/s. From a distance and edge-on to the Solar system, the Sun would appear to approach at a speed of 12.5 m/s and then recede at the same speed, with a periodicity of 11 years. That's only about 45 km/h, or 30 mph (miles per hour), less than half the speed of an approaching train or ambulance. It's not very fast, but with the precision of Doppler technology, it's measurable.

The first confirmed use of the radial-velocity technique to detect a planet orbiting a distant star was in 1995. The star was 51 Pegasi, and the planet was named 51 Pegasi B. The speed of 51 Pegasi was measured at 60 m/s. The orbital period was 4.2 days. With some assumptions about the star's mass, this information on its periodic radial velocity can be used to estimate the mass of its orbiting companion. In fact, the calculation can only reveal the minimum value of the companion's mass. This uncertainty is a result of not knowing the orientation of the orbit. We may not be viewing it exactly edge-on. All the information we have to go on, after all, is in the periodic spectral shifts. Nonetheless, this estimate of the mass is enough to determine whether the companion was another star that is too dim to see or a planet. The calculated mass of 51 Pegasi B was, at a minimum, one-half the mass of Jupiter.

If the exoplanet orbits in a plane that is perpendicular to our line of sight there will be no radial motion of the star and hence no Doppler shift to measure. The star moves slightly across the field of view, and in some rare cases it is possible to detect. This is called the astrometric technique for detecting exoplanets. Again it is the periodic motion that is characteristic of an orbit, but it's not the velocity that is being measured this time, it's the actual change of position. Such changes are minute. The Sun, don't forget, changes position only by a distance roughly equal to its own diameter. Measurements to detect this kind of proper motion must be extremely precise and are only feasible for nearby stars. Furthermore, the success of this technique requires observing the same star over a long period of time. It is an inefficient technique, but theoretically straightforward. The earliest claims to astrometric detection of exoplanets were controversial and uncorroborated, and this technique is still very limited.

Since 2009, the work-horse in exoplanet detection has been the Kepler orbiting telescope. Monitoring thousands of stars over long periods of time, Kepler alerts astronomers to the event of an individual star suddenly dimming. This is the photo-metric technique. It has nothing to do with the motion of the star, only its brightness. A subtle decrease in brightness might be caused by a dark object, a planet, say, passing in front of the star. The decrease in brightness of a Sun-like star being occulted by a Jupiter-like planet would be minimal, roughly 1%, but this is measurable.

The photometric technique is vulnerable to false positives, since other things besides passing planets can block a bit of the light from a star. Kepler is responsible for most of the "candidate" exoplanets, but independent corroboration is required for confirmation. Measuring just how much the starlight is reduced gives an esti-mate of the planet's size. Long-term observation can reveal the period. It would take a year's worth of data if the planet had an Earth-like orbit, 11 years if it was like Jupiter. Photometric measurements do not provide information on the mass of the planet, but they do indicate that the orbit is almost exactly edge-on to our line of sight. If there are also radial-velocity data for this same star, the confirmation that the Doppler-measured speed is the actual speed of the star allows a more precise determination of the mass of the planet. In this way the different methods, attending to different properties of the star, are complementary.

Determining a planet's mass is a challenge. It's also an important piece of information for assessing the potential for life. The Earth is just the right temperature to sustain life, and it's the right composition and density; it's solid. The temperature is a consequence of the size and intensity of the Sun, and the distance between the Sun and the Earth. So the type of star and radius of planetary orbit are important. Neither of these depends on, or is indicated by, the mass of the planet. But the density is. If you know the size, the diameter of the planet, and you know its mass, you get the density. Density distinguishes a planet from a small star, a brown dwarf, for example, and is one of the conditions for the possibility of life.

The mass of the planet would also be useful information to test a theory of gravity. All theories of gravity since Newton say that two objects affect each other in a way dependent on the mass of each. To see if such a theory is true would require situations in which both masses are known. We can measure the mass of something on the Earth, but to claim *universal* gravitation there must also be celestial situations of known masses. Exoplanetary systems might be just the thing, as long as the mass of both star and planet can be known.

Finding unseen massive objects from their gravitational effects on things that can be seen is a recurring tool of discovery in astrophysics. Finding Neptune was one case. From the motion of Uranus the existence of Neptune was inferred. With the precision of Newtonian theory, the position and mass of Neptune was calculated. This is very much the same procedure as detecting an exoplanet, but no one declared confirmation of the discovery of Neptune until the actual direct observation with a telescope. Neptune was not detected by looking at Uranus, but 51 Pegasi B was detected by looking at the star 51 Pegasi. What's the difference? And what about Vulcan? Here is another case of carefully measuring the movement of one thing that can be seen, Mercury, and calculating the mass of another that cannot. Why is this not described as detecting Vulcan?

The answer seems obvious in the case of Vulcan. That was not detecting a planet because there was no planet there to detect. But that answer is not helpful, or scientific. It depends on hindsight and evaluating the procedure by knowing the answer. Scientists are generally in the situation of having to figure out the answer by using an independently evaluated procedure. And the criteria of confirmation must be applicable at the time. That's scientific reasoning. So the question is, what allows scientists *now* to say that 51 Pegasi B and almost two thousand other exoplanets have been detected, but disallowed scientists *then* from saying Vulcan – or Neptune before 1846 – had been detected? One more example can be added for context. **Dark matter** deserves an entire chapter to itself, but it's worth noting here that it is another case of unseen mass – it's dark, after all – detected by its gravitational effect on what can be seen. There's a lot of it, and it is by some accounts a different kind of matter than anything we experience. We'll deal with it in the next chapter.

It's clear that using gravitational effects to find and measure an unseen distribution of mass is both important and useful. But it's also something of a conceptual challenge. Using the radial-velocity technique to detect an exoplanet, how can we determine the mass of the planet? The basic empirical data are all kinematics, the period of the orbit and the orbital speed of the star. That's not enough information to calculate the mass of the planet. The kinematics underdetermines the dynamic property, the mass. In fact, there is not enough information here to pin down a value of the distance between the star and the planet, and that's key to the possibility of life. What more information is needed to get the planet's mass and orbital radius, and where does that information come from?

First, you have to know the mass of the star. With that, using what we called the orbit equation in Chapter 3 (Equation (3.7)), the planet's orbital radius can be found. And then it's a simple application of the law of conservation of momentum to get the mass of the planet. Some simplifying assumptions must be made along the way.

Here are the details. The binary system of star and planet is shown in Figure 13.1. We'll set up the calculation as if the orbit is exactly edge-on to our observation. That way the Doppler-shift measurement gives the actual value of v_s, the orbital velocity of the star, and not just the component of the velocity along our line of sight. We'll also assume, as Newton did in his original derivation of the law of universal gravitation, that the orbits are circular. Planet and star each orbit the center of mass, the barycenter, at radius r_p and r_s, respectively. This puts the distance between the two at $r_p + r_s$.

From details of the electromagnetic spectrum emitted by the star, astronomers can estimate its luminosity, and from this they can estimate its mass. This is facilitated by an extensive cataloging of stars and noting correlations between one property and another. A star's spectral characteristics are directly observable. So is its apparent brightness. If you also know the distance, apparent brightness can be converted to absolute magnitude or luminosity. With this conversion for stars that are close enough to judge distance by parallax, a distinct correlation shows up between spectral type and luminosity. This is plotted on a Hertzsprung–Russell diagram, allowing an easy link between measured spectrum and luminosity. Careful cataloging has also revealed correlation between the luminosity and mass of a star. Astronomers use an equation called, naturally enough, the mass–luminosity

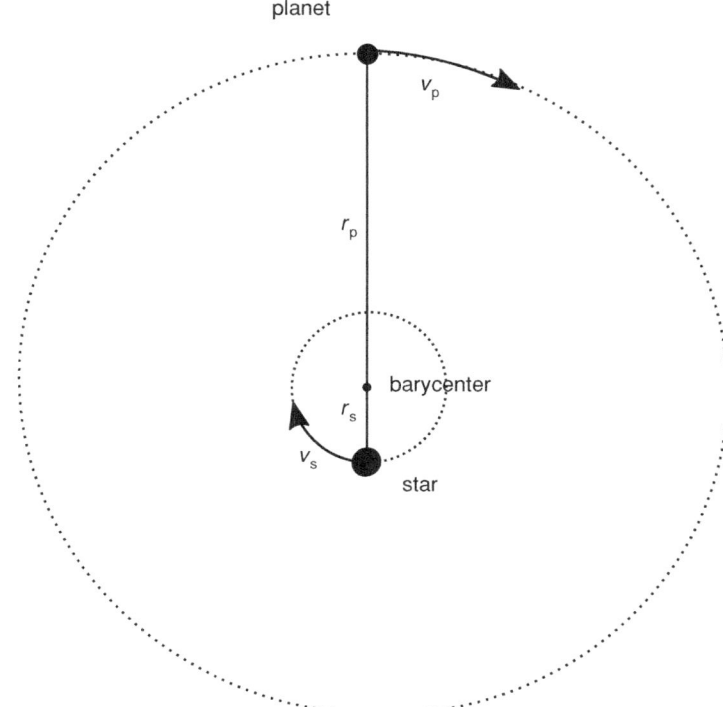

Figure 13.1. A binary system of a star and a planet. Both the star and planet orbit the center of mass between them, the barycenter.

relationship. So, in the exoplanet system, we get m_s, the mass of the star from the spectral profile of the star.

But where does the mass–luminosity relationship come from? We have to ask, since one of the reasons for this detailed analysis is to see how measuring the mass of a celestial object depends on a theory of gravity. The empirical basis of the mass–luminosity relation depends on binary stars that are close enough to know their distance by stellar parallax. Two visible stars at known distance and with measurable orbital radii and period allow calculation of the mass of each. It's an application of Kepler's third law, as we're about to do for the exoplanet, but with more data. Knowing all the kinematics, since in a binary star both orbiting objects are seen, determines the masses, as long as we know how gravity works. So nearby stellar binaries offer data on luminosity and mass, and a catalog of these reveals the correlation, the mass–luminosity relationship.

Back to the exoplanet. We now know the mass of the star, and from this we can calculate r_p the radius of the planet's orbit and m_p the mass of the planet. Assuming, again as Newton did, that the mass of the star is much greater than that of the planet and consequently r_p is much greater than r_s, Kepler's third law becomes the following:

$$P^2 / r_p{}^3 = 4\pi^2 / Gm_s \qquad (13.1)$$

Here, P is the period of the planetary orbit, and this is exactly the same as the period of the stellar orbit. Solve Equation (13.1) for r_p and use that to find the planet's orbital velocity v_p. Velocity is simply distance travelled, the circumference of the orbit, per time, the period of the orbit:

$$v_p = 2\pi r_p / P \qquad (13.2)$$

One more step and we'll have the mass of the planet. As the planet and star orbit the center of mass, the barycenter, that point does not move. So the total momentum of the system doesn't change. Use this law of conservation of momentum to equate the star's momentum to that of the planet:

$$m_s v_s = m_p v_p \qquad (13.3)$$

Solve for m_p.

We wouldn't have made it without a theory of gravity. We wouldn't have even started the process that brought us to "solve for m_p" without a theory of gravity. It was in at every step, from the empirical mass–luminosity relationship to Kepler's third law. There was also the subtle assumption that the gravitational constant G has the same value on the Earth, where it has been measured, as in the celestial binary system, where it has been applied in the calculation. There is no way to measure G in the celestial context without a theory of gravity, just as there is no way to measure mass without a theory of gravity. The "universal" in universal gravitation comes from the successful application of the theory here as well as there, successful in the sense of maintaining consistency. It does not come from an empirical foundation of independent data.

In fact, a lot of theories come together on the discovery and description of exoplanets, and their cooperation is a key aspect of their confirmation. There is theory on how planets form. This gives the plausible context for finding exoplanets. There is theory on how stars shine, and this corroborates the mass–luminosity relationship. There is our understanding of the Doppler shift and signature spectra of planets. And of course there is a theory of gravity. None of these is based on a foundation of purely empirical data. It's the coherent collaboration in using the theories to interpret observations that gives them a measure of confirmation.

So, determining the mass of the exoplanet is a matter of cooperation between observations and interpretive theories. Even determining that the planet is there, the basic detection, comes down to confirmation by corroboration. The so-called candidate exoplanets put on that list by the Kepler telescope are detected by photo-metric means. These are promoted to being confirmed if radial velocity measure-ments of the same star are consistent with the hypothesized planet. In other cases the order is reversed. Regardless of the order of measurements, it is the multi-method agreement that astronomers describe as confirmation. The confirmation is not in one method or the other. Neither method provides the more foundational or more direct observation than the other. The confirmation is in their agreement. Indepen-dent sources reduce the chance of misinterpreting the results or being fooled by an artifact of instrumentation or some non-planetary cause of the star's behavior.

Apply this standard of confirmation to the historical cases of Neptune and Vul-can. In neither case was there the kind of multi-method corroboration that there is for exoplanets. This is at least part of the good reason why detecting unusual motion of Uranus or Mercury lacked the empirical status to declare these discov-eries of new planets.

Another important difference between then and now is the theoretical context of the evidence. The evidential results of looking for exoplanets fit coherently into a larger theoretical understanding of nature. Other planetary systems are to be expected. Observations of proto-planetary disks give this expectation empir-ical reinforcement. So, evidence indicating the existence of actual exoplanets gains credibility by fitting into this theoretical scheme. The theoretical context of Nep-tune and Vulcan was significantly different. There was no clear understanding as to how planets were formed, and no expectation as to the number of planets in the Solar system. So there was no clear theoretical reason to expect Neptune or Vul-can. The observations of the motions of Uranus or Mercury did not have such an accommodating theoretical niche to fit as do current measurements of wobbling and dimming stars. Without the theoretical sponsorship, the former were not so immediately evidence of other planets as to call them detection.

This comparison between detecting exoplanets and not detecting Neptune or Vulcan is meant to highlight the essential structure of scientific method. It's the "tangled web of theory and experiment" described by Weinberg. Neither source of information, theory nor experiment, is foundational. Their credibility is in their cooperation.

There is one more thing about the detection of exoplanets that points to the insep-arability of theory and observation, another strand in the tangled web. Whichever technique is used, there is reliable information tracking from object to image. We don't have to actually see the planet to detect it. The information does not have to be carried by light, as long as we understand how the information gets from the planet to us. Using the radial velocity technique requires a basic understand-ing of the gravitational effect of a planet on the star we can see. It requires an understanding of optics and spectral analysis and Doppler shifts. And it requires an

understanding of the difference between planets and stars. All of these requirements are met at a satisfactory level. We do not understand everything about all of
these things. In particular, theories about stars and planets are imprecise and corri-
gible. But the basics are secure and sufficient to track the information in detecting
exoplanets.

Again, compare the case of exoplanets to those of Neptune and Vulcan, this
time by the criterion of reliable information tracking. In the historical cases, the
interaction between what was observed and the unobserved cause was gravitational.
But the only account of gravity at the time was Newton's law, and the reliability
of that theory was part of the question. In other words, there was much less reason
to trust the theoretical account of information from object to image in the cases of
Neptune and Vulcan than in the case of exoplanets.

It's not that the theories describing the flow of information have to be true; they
only have to be reliable. There's a difference. A key theoretical player in the radial-
velocity and astrometric methods is an account of the gravitational interaction. That
was the reason for our interest in exoplanets to begin with. But it's the Newtonian
theory that is used. That will do because planets are pretty slow and the gravitational
fields are pretty weak, so relativistic effects are negligible. Strictly speaking, this
part of the theoretical tracking of information is false, but reliable. In this case, and
in general, accurate information about unobservables, or indirectly observables,
can be accessed with the help of false theoretical support.

A theory like the Newtonian theory of gravity, or the general theory of relativity,
can be useful in discovering other things about nature, even if it's not altogether
true. This works in applying gravitational lensing. Gravity bends light, and you
can use this, in the same way the light-bending abilities of a glass lens are used, to
see things in outer space.

Evelyn Gates calls this "Einstein's Telescope" (Gates, 2009). It's an evocative
description, but a little misleading. Since the phenomenon of lensing is not unique
to Einstein's theory of gravity, it's not just Einstein's telescope. Any theory in which
massive objects bend light will result in lensing and hence a "telescope." The exact
mechanism of lensing and the quantitative relation between mass and the trajectory
of light will differ theory to theory, so the details on tracking the flow of information
through the telescope will differ. But there could be a Newton's telescope as well.
At some point we'll have to ask whether it's possible to tell whose telescope we're
using, Newton's or Einstein's.

Another clarification to Einstein's telescope is that it is generally the lens, the
focusing element, that is the object of astronomical interest, not the more distant
source of light. The image of a galaxy or quasar is altered under the effect of lensing
by the mass, maybe unseen mass as of a black hole or dark matter, and this is
evidence of the mass. There is a similar situation in geophysics in techniques for

detecting and analyzing properties of the core of the Earth. Earthquakes generate seismic waves that are reflected and refracted by the core, to be detected at very distant, perhaps opposite, locations on the surface of the Earth. The core acts as a lens by bending the incident seismic waves, and careful analysis of the waves that arrive on the surface provides information on the characteristics of the core. The lens itself is the specimen.

One more thing about the reference to Einstein's telescope. A telescope, an optical telescope, always requires more than one lens, carefully aligned. Galileo used two lenses, carefully positioned along the tube. Gravitational lensing, Einstein's telescope, gets just one lens.

It's fair to call it a lens, since lenses work by bending light, and gravity bends light. It's also fair to point out that you can use a lens to great advantage without knowing how or why it bends light. Galileo had no real understanding of optics. Newton had a clear understanding, but it was wrong. He theorized that light was a stream of particles. The particles speeded up at the interface between air and glass, and that's what made a ray of light bend in, toward the glass. He had a very serviceable optical theory, and used it to produce very effective telescopes, but it was fundamentally different from the wave theory used today. Fair warning.

Lenses come in two varieties, converging and diverging. The typical converging lens is convex, thicker in the middle than at the edges, like a lentil. A diverging lens is concave, narrow in the middle, flaring out on the edges. A converging lens draws light rays closer together. A diverging lens spreads them apart.

A converging lens can produce a real image in which the rays of light from an object are brought together and focused to produce an actual glowing reproduction of the object. You need a real image in a camera so that the light from the subject is focused onto the film or the image sensor. A converging lens can also form a virtual image if the object is closer to the lens than the focal length. The rays of light emerging from the object are bent inward by the lens, but not enough to get them to intersect. There is no illuminated reproduction. Instead, an observer's eye traces the rays back to where they *seem* to be coming from, and that is where the object appears to be. A single-lens magnifying glass works this way, since the virtual image is larger than the object itself. A diverging lens by itself can only produce a virtual image. It never focuses the rays emerging from an object into a real image.

It might seem that a gravitational lens will always be converging. Gravity, after all, is only attractive, never repulsive. And a glance back at Figure 10.8 shows the rays of light bending exactly as they would going through a convex lens. There are cases of converging gravitational lensing, but there are also situations in which the lens is diverging. Rays of light are always bent in toward a massive object, but close-in rays are bent more sharply than rays that are further out. Follow the two rays of light from the star in Figure 13.2 and note that they are diverging from each

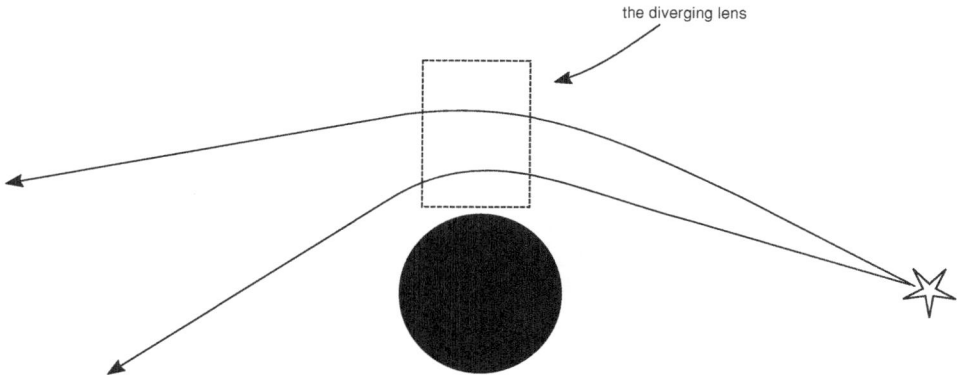

Figure 13.2. Gravitational lensing creating a diverging lens. The two light rays from the star are bent toward the black hole as they pass by, but the closer ray is bent more than the more-distant ray. The result is that the rays are diverging from each other more after they pass the black hole than they were when they first left the star. This is the effect of a diverging lens.

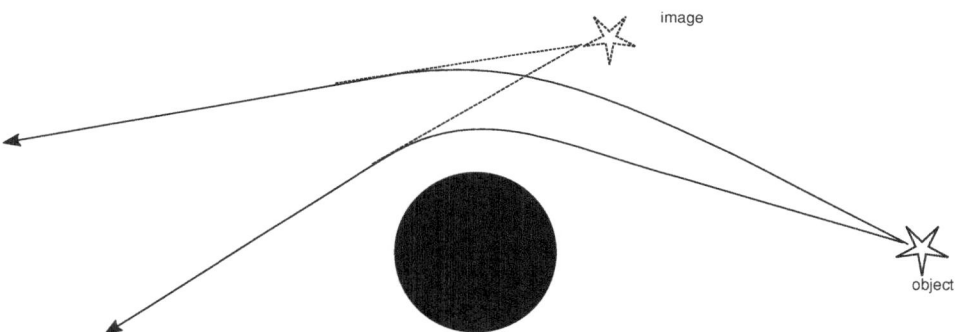

Figure 13.3. The shifted image produced by a diverging lens. The image of the star is found by tracing the light rays straight back to where they would converge. This image is shifted out from behind the black hole.

other more after they pass the black hole than they were before. The lens is not the whole black hole but only the edge, and it's a diverging lens.

Like any diverging lens, this one produces a virtual image. Trace the observed rays back along straight lines, dashed lines in Figure 13.3, and they meet at the image. This is where the star will appear to be, shifted out a little bit from behind the black hole. It will also appear closer and brighter than it would without the lensing effect.

All of these effects, the shifted, closer, and brighter image, are in comparison to the unlensed star. But the black hole is not going to move out of the way any time soon. We'll never see the star without the lens. It's not like the 1919 solar eclipse

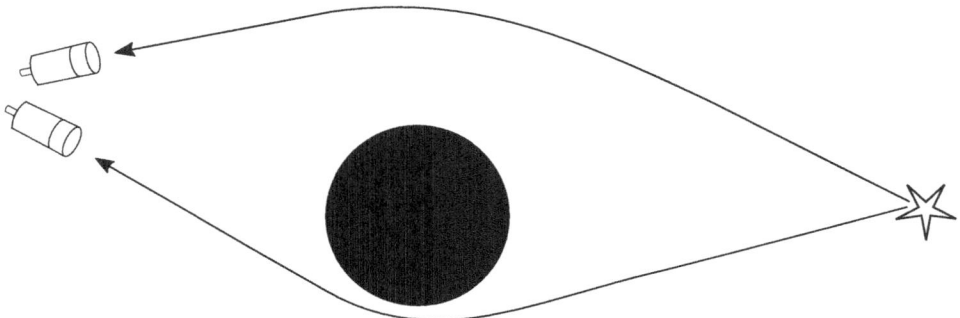

Figure 13.4. Two images of the same object formed by gravitational lensing. The star is seen by pointing the telescope in the direction of either of the light rays shown. The result is that the star appears at two different angular positions in the sky. It appears at two separated places in the sky.

expedition where, in six months, the lens, in that case the Sun, would be nowhere near the object stars. So, in the situation at hand there is really no way to tell that any lensing has happened, let alone how much the light has been bent. We could call this a way of detecting the black hole, by its lensing effect on the starlight, but without knowing the actual position and brightness of the star there really are no data.

There are situations, though, in which lensing is verifiable. Figure 13.4 shows a case of multiple images of a single object, the result of lensing. Look in one direction in the sky, along one of the light rays shown, and you'll see the object. Look in another direction, determined by the details of the lensing, and there is exactly the same object, clearly identifiable by it spectral characteristics. If the lensing mass is evenly spherical and the bright object, the lens, and the Earth are all in a line, there will be not just two images but a ring. It's called an Einstein ring, and they have been observed.

The amount of bending shown in Figure 13.4 is an exaggeration. We could calculate the angular size of the ring for a typical case, knowing the lensing mass and the distances from mass to bright object and mass to ourselves. The calculation, of course, is based on a theory of gravity, and the result will be different for Einstein's theory and Newton's. So how do we know if we're looking at an Einstein ring or a Newton ring? Never mind the challenges of measuring the distances to the lens and star, but don't forget that the lens may be a back hole and hence invisible, and that the star is visible only as distorted by the lens. The real unknown is the mass of the lens. There is no measure of celestial mass independent of a theory of gravity.

Since an understanding of gravity and a specific theory of gravity are everywhere in the details of lensing, using lensing to measure the details is going to be limited. Finding multiple images or a ring of a single celestial object is indeed evidence of lensing. That means there must be some massive object, or collection of objects, doing the lensing. In this way, lensing and the theory of gravity can be used to "probe the mass distribution" (Schutz, 2003, p. 331) of the universe. But determining the *amount* of mass will require assuming the accuracy of a particular theory of gravity. This will be no independent test of the theory, and no verification of the power of the telescope. Galileo could at least use his telescope on distant ships at sea to demonstrate that, when the ship came to port, it looked in real life just as it did through the telescope. There is no opportunity to compare telescopic image to the object itself in the case of gravitational lensing. It's dependent on theory all the way. This will be of particular concern in detecting dark matter.

Using a theory as a means to discover other things, things unrelated to the concerns of the theory itself, may be a good way to show the theory is true, or likely to be true. You take it for granted, apply it, and that way you show it works without really trying. The special theory of relativity was used to design the Stanford linear accelerator, and the accelerator works. Doctors use a PET scan, based on the amazing notion of anti-matter – the P stands for positron, the anti-twin of an electron – and the scanner works. Diseases are diagnosed with the guidance of the image and patients are cured. How could all of this happen if there were no positrons, if the theory was false?

But of course, it could. There is again the cautionary tale of caloric. It was caloric theory, and the plumbing of caloric fluid, that was used in the invention of steam engines. Caloric theory started the industrial revolution. It worked wonders, but it was false. There is no such thing as caloric fluid.

In the logical jargon of methodology, successful use underdetermines theory. This is the basic idea of instrumentalism. Using a theory to successfully get things done does not indicate that the theory is true; it indicates that the theory can be used to get things done, and nothing more. Exoplanets can be detected using Newtonian gravity or the theory of relativity. Black holes can be discovered by gravitational lensing, again using either theory. In the case of planets, either theory works because they give the same results in the circumstances and within the precision of the data. With lensing the results are different, but outside of the Solar system there is no way to tell which results you are dealing with.

The logic of using a theory is the same as the logic of testing and explanation, and that's the source of the underdetermination. It goes: if the theory is true then we can make such and such happen. If there is caloric fluid it will flow from hot things to cold and make wheels turn as flowing water turns a water wheel. Things spin, but to claim that this demonstrates the existence of caloric is to commit the fallacy of affirming the consequent.

Successful use of a theory can claim some evidential advantage over targeted testing. Part of the advantage is personal, an indifference on the part of using the theory, and part is logical, an independence between the theory and the phenomenon it's used to produce or investigate. Neither doctor nor patient cares if there really are positrons. They might not even know what the P is in a PET scan. So there is no danger of interpreting any of the data under the influence of the anti-matter theory. This prevents the sort of adjustment of observation to theory we saw in the previous chapter when the theory of relativity was being tested.

Using a theory doesn't prove that it's true, but it does tie the theory in yet another way into the tangled web. And the connections can be to very different kinds of phenomena than are the obvious subject matter of the theory, novel phenomena. This is a kind of independent corroboration. It's a little tricky, though, with a theory of gravity, at least when using it to explore the universe and map the distribution of mass. There is less opportunity for independence. To test the theory you need to know the amount of mass in a system, but to measure the mass you need a reliable theory of gravity. There is virtue in getting all the parameters, kinematic and dynamic, into a consistent account, but there is no outside-the-theory corroboration.

This is particularly challenging in the detection of dark matter.

14

Dark Matter

Nature is not as it appears; that's the challenge for science. In some cases the reality is revealed just by changing how we look. What appears to be a single bright star turns out to be two stars, if you look through a telescope. The Milky Way, it turns out, is not so much milky as it is grainy with stars, again through a telescope. But there are some things that no manner of magnification will reveal. There is, for example, dark matter.

It's not clear what the dark matter is, whether it's just more of what we already know to exist, things like black holes and brown-dwarf stars, or it's an entirely new kind of stuff. What is clear is just how much of it there is, that is, what percent of the mass in the universe is dark matter. The number is reported with confidence, and it's remarkable, not just the amount but also the precision and tone of certainty. The current accounting of mass in the universe puts dark matter at 27%. Only 5% of the universal mass is what you would call normal matter, the usual stuff we can see, made of protons, neutrons, electrons, and the like. The rest is invested in what's called dark energy, a new and important aspect of cosmology that we will not deal with. We can't ignore the dark matter; it's too important to our interest in both gravity and scientific method.

The great quantity of dark matter, and the willingness to bet on it, might be troubling to a theorist of principle such as Einstein. The number is entirely adjustable, like properties of a Ptolemaic epicycle, to accommodate the latest astronomical measurements. How much dark matter is there? Well, how much do you need? There is no theoretical constraint on this, so name a number that takes care of any discrepancy between what you observe and what you expect. It's convenient, especially with the dark matter being dark. With an unlimited supply, a hypothesis of dark matter may evade meaningful testing. It may not be falsifiable. We need to look carefully at the details of detection, and the interpretive role of the general theory of relativity. It's a tricky situation, looking for dark matter. Since we don't know exactly what it is, it's a case of looking for something while at the same time figuring out what it looks like.

The idea of dark matter has been around since the 1930s. The Swiss astronomer Fritz Zwicky in 1933 hypothesized a significant amount of invisible mass in the Coma cluster of galaxies. He called it *dunkle Materie* (van den Bergh, 1999, p. 657). Like any seriously considered idea in science, this was not a whim or baseless speculation. There was some empirical motivation, some evidence, though very indirect evidence. It was not from observing the *dunkle Materie* itself, as seems to be ruled out in principle, but in observing properties of other things. Zwicky measured velocities of galaxies in the Coma cluster and found average values much higher than expected. More mass was needed to account for the motion, so Zwicky suggested there was just that, more mass.

The invisible mass was not anything exotic, nothing other than the already known kinds of matter. It was suggested to be things like brown-dwarf stars and cold inter-stellar gases. You might call it dim matter rather than dark matter, since, if we were close enough to it, it would be observable. The argument for its existence started with an assumption of the basics of the current theories of gravity and mechanics. We have seen precedents for this kind of theoretical maneuver, proposing new matter to bring existing theory in line with anomalous data. Neptune, Vulcan, and exoplanets all follow this pattern. It also shows up in other parts of physics, particle physics for example.

Neutrinos were theoretically discovered in 1930, a new and elusive form of matter introduced to explain surprising details of radioactive decay. The energy spectrum of beta decay was observed to be continuous, presenting a challenge to quantum theory. The continuous spectrum of beta decay was unlike the discrete spectra observed in both alpha and gamma radiation. Wolfgang Pauli suggested that there was an unseen particle produced in the beta decay, a particle that carried away some of the energy. It might well have been called "light matter," given its zero mass, but in 1933, Enrico Fermi named it the neutrino, the little neutral one. The hypothesized matter was given the appropriate properties to save both the law of conservation of energy and the theory of quantum mechanics.

The discovery of Neptune, as we saw in Chapter 8, is an exemplar for the benefits of waiting out anomalous data. An eighth planet, as yet unseen, would explain the evidence and save the theory. In this case the hypothesized matter, the planet Neptune, was not so much dark as it was hiding. All that was required was knowing where to look.

Neptune would be an encouraging example for a dark-matter theorist. Vulcan would not. Vulcan was hypothesized to save Newtonian gravity, but Vulcan was never found, and in this case it was the basic theory of gravity that was changed.

Since its inception in 1933, the hypothesis of dark matter has benefited from other indirect evidence beyond accounting for galactic speeds in clusters. For example, the explanation of the formation of galaxies and rotational speeds within galaxies requires more mass than is present in the visible matter. Adding the mass in dark matter provides the necessary dynamic mechanism. This evidence, like the evidence used by Zwicky, is indirect in the sense that it is the result of an inference from an explanation of what is observed in something else, something other than the dark matter itself.

Recently, gravitational lensing has been added to techniques for finding evidence for dark matter. It's regarded as almost as good as seeing Neptune through a telescope, just using Einstein's telescope. A 2006 article in *The Astrophysical Journal* claims "A Direct Empirical Proof of the Existence of Dark Matter" (Clowe *et al.*, 2006, p. L109). Summarizing the proof and endorsing the optimism, Evalyn Gates says, "The evidence is clear, . . . We now know that dark matter of some form exists" (Gates, 2009, p. 194).

The evidence may be "clear" and "direct," but it still requires some sophisticated theory to link what is observed to what the observation is evidence of. A significant component of that is a theory of gravity. This raises some methodological questions and challenges. What makes the evidence direct rather than indirect, and is this distinction of any real importance? Why not question, rather than use, the current theory of gravity, the general theory of relativity? This seems to be a clear case of underdetermination of theory by evidence, so why, particularly in light of the case of Vulcan, is the conclusion not a disjunction of the form, either there is dark matter or the theory of gravity we're using to interpret the evidence is wrong?

The general theory of relativity has been tested, and the results are reasonably positive. This seems adequate reason to then use the theory as the interpretive tool to explore the universe. But the testing has been all very close to home, within the Solar system. Dark matter is in systems very much larger and very far away. Will our local, small-scale theory of gravity apply accurately in the expanse and enormity of a cluster of galaxies? Two things are different between the circumstances in which the theory has been tested and those in which it is used to detect dark matter. One is that the masses and distances between them are much larger there than they are here. The other is simply that that's there; this is here. This second difference may not make a difference, if gravitation is truly universal. The laws of gravity are the same here as they are there. But that's just an assumption, maybe a good and safe assumption, but untestable if the theory is used to interpret the evidence. It follows the **Copernican Principle** that our situation in the universe is entirely ordinary and the laws are commonplace.

Adopting that principle, at least as a working hypothesis, let's see how dark matter is detected, indirectly and directly.

To put things in perspective, here are some of the numbers. The distance to the Sun is 1.5×10^{11} m, or 1 AU, one Astronomical Unit. The Solar system is about 60 AU across. The distance to the nearest star is 5×10^{16} m (4.5 light years), or about 300 000 AU. That's the *nearest* star. The Milky Way galaxy is roughly 60 000 light years across. Distances between galaxies run at about 1 000 000 light years. That's on the order of 10^{11} AU. So applying a theory of gravity at galactic distances will bring it at least 11 orders of magnitude outside its testing zone. The masses it encounters will be much larger as well. The Sun has a mass of 2×10^{30} kg, or $1 M_S$, one solar mass. The mass of the Milky Way is $10^{11} M_S$. It's a pretty typical galaxy, as the Copernican Principle would expect.

We've asked this before, but how is the mass of a distant astronomical object measured? In particular, how is the mass of a galaxy or a cluster of galaxies measured? The answer gets us to dark matter.

One way to measure the mass is to use the mass–luminosity relationship as we did with the individual star holding an exoplanet in orbit. With a galaxy, measure the total production of electromagnetic radiation, that is, not just light but radio waves, ultra-violet, and so on. Then, with a theoretical understanding of how the radiation is produced, together with the empirical correlations between luminosity and mass in stars, calculate the total mass of the galaxy. Call this the luminous mass, or call it the mass of the visible matter.

If the galaxy is spinning, as ours is, there is a second way to measure its total mass. Again drawing on the similarity to a star–exoplanet binary, or to the Solar system, knowing the kinematics of rotation provides information on the amount of mass needed to hold things in orbit. A spinning galaxy is a huge system of stars in orbit around the collective center of mass, just as the Solar system is a group of planets spinning around the collective center of mass. There is a correlation between the radius of a planet's orbit and its orbital speed. It represents the balance between centripetal force and the force of gravity. It is exactly what we called the orbit equation, Equation (3.7) in Chapter 3:

$$v^2 = GM/r \qquad (14.1)$$

In the case of the Solar system, M is M_S, the mass of the Sun. The kinematic data on the planets, namely their distance r from the Sun and their orbital speed v, are enough to solve for M. Plotting these data on a graph of v as a function of r is called a **rotation curve**. The rotation curve of the Solar system is shown in Figure 14.1. The shape of the curve shows that v is proportional to the inverse of the square-root of r. The fact that the measured properties of r and v for each planet put them exactly on the theoretical curve gives some confirmation of the basic theory of gravity, that the force varies inversely as the square of the distance, the inverse-square relation.

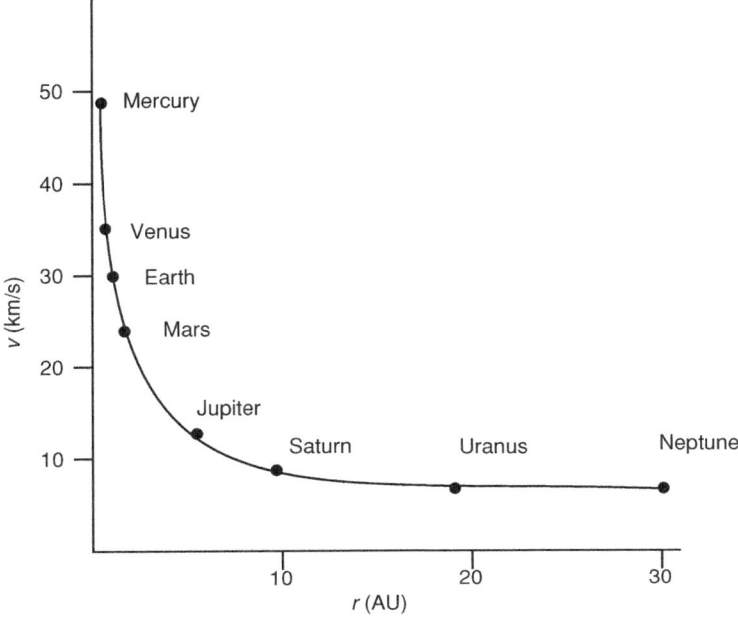

Figure 14.1. The rotation curve for the Solar system. The solid curve shows the calculated values of the speed v of an object orbiting the Sun, as a function of its distance r from the Sun. This a plot of the orbit equation. The measured values of speed and distance from the Sun for each of the planets in the Solar system are labeled.

The details of the shape, the constant of proportionality, allow for the determination of the mass M.

But that's the Solar system. A galaxy is more complicated. The orbital speed of a planet is usually measured by knowing the period of the orbit and the radius. The period of rotation in a spinning galaxy is much too long to measure. It takes us 250 million years to get once around in our galaxy, for example. But if the galaxy is spinning edge-on, the orbital speeds of the stars can be measured by their Doppler shift. So data on both v and r are available. The real complication is in the mathematical relation between v and r. The orbit equation, the basic $v \propto 1/\sqrt{r}$, was derived by assuming that all of the gravitational force on the orbiting body came from a single mass at the center of the system. We did each planet one at a time, considering only the central mass of the Sun, ignoring any effect of the other planets. That works when the planets are few and not very big. But a galaxy is made of stars, lots of them, and they are big. There is no central dominating mass.

In the spinning galaxy, the average star, somewhere in the mix, has massive stars inside its orbit, and massive stars outside, further from the center. For stars near the outer edge of the galaxy the orbit equation is the accurate correlation between v, r,

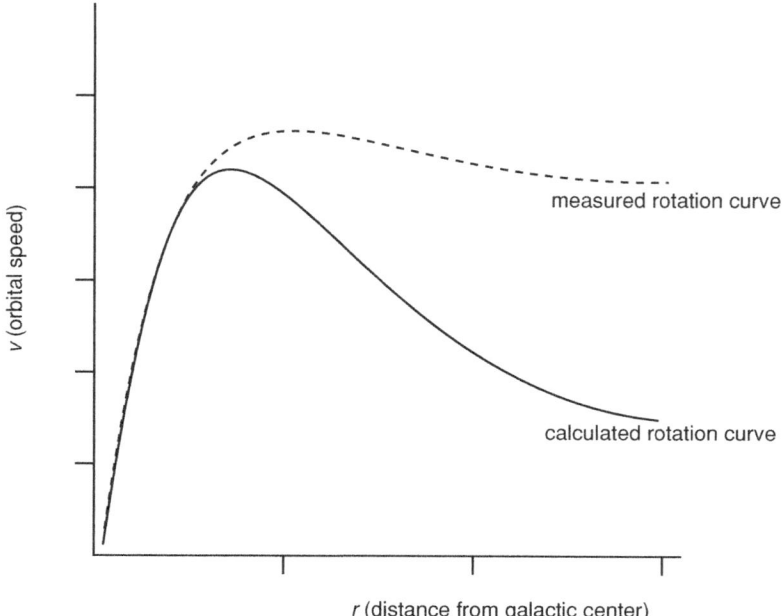

v (orbital speed)

measured rotation curve

calculated rotation curve

r (distance from galactic center)

Figure 14.2. A rotation curve for a rotating galaxy. The curve depends on the amount and distribution of mass in the galaxy. The solid curve shows the calculated relation between orbital speed and distance from galactic center, using the visible objects to determine the information on mass distribution. The dashed curve shows actual measured values of orbital speeds and distances from the galactic center. The dashed curve is consistent with there being more mass in the galaxy than is visible.

and the mass of the entire galaxy *M*. All the mass within the star's orbit acts as if it was all concentrated at the center. But, for stars closer in, the value of *M* in the orbit equation decreases; it's not the total mass of the galaxy but only the mass within the orbit. The stars outside the orbit have no gravitational effect; their forces cancel. The result is that the orbit equation has to be revised, making *M* itself a function of *r*, decreasing as *r* decreases. Less mass means less centripetal force and consequently less centripetal acceleration. Close to the center of the galaxy, where there is not much central mass at all, the orbital speed will be slower than the simple planetary-system orbit equation would predict. As *r* gets close to zero, so does *v*.

It takes an act of calculus to derive the mathematical expression for this galactic rotation curve, but we can picture what the curve must look like. At large values of *r* the curve matches the original orbit equation with $v \propto 1/\sqrt{r}$. At $r = 0$ the curve goes to $v = 0$. Figure 14.2 fills in the points in between and shows the calculated rotation curve with the solid line. The curve incorporates the information on how much mass there is in the galaxy, and how it is distributed around the center of rotation. This information comes from the visible objects.

As in the Solar system, these parameters can be measured. But this time the values do not follow the predicted curve. A typical array of actual values is plotted as a dashed line on Figure 14.2. For small values of r, the speeds are as predicted, but for large values of r, further from the center of the spinning galaxy, the speeds are much higher than predicted. Higher speeds require more centripetal force to hold things in orbit, and that means more mass. There must be more mass than we see. There must be dark matter.

This is the nature of the indirect evidence for dark matter. More mass is needed to account for the gravitational dynamics to hold the kinematics together. A similar analysis applies even if the system is not rotating and there is no orbit equation or rotation curve to consider. In those cases, for example in a cluster of galaxies, it's a comparison between the gravitational potential energy and observed kinetic energy of the galaxies that shows there is missing mass. The speeds of galaxies are measured and reported as a velocity dispersion, the spread of values with respect to the mean. A theory of gravity is used to link potential energy to mass distribution, and in this case either Newtonian theory or the general theory of relativity will do, as their results agree to within the limits of measurement. The dynamic properties of the mass distribution, and the resulting gravitational potential energy are linked to the kinematics of speeds. As in the spinning galaxies, there is a mismatch between the theoretically predicted velocity dispersion and the empirically measured velocity dispersion.

There are three things that could be adjusted to restore agreement between theory and observation in these cases. It could be that there is more mass in the system than is visible; this is the dark-matter solution. Or it could be that the theory of gravity used to draw the rotation curve or velocity dispersion is wrong. Or it could be that the measured values of speeds of stars and galaxies are wrong. This third option is pretty much not an option at all, since the kinematics of speed is the most directly empirical component in the network. But the other two factors, the amount of mass and the theory of gravity, are dynamics. Neither is directly empirical.

The logic in this case is identical to the logic of testing. A theory makes a prediction that does not match the measured results. The careful, that is, logical, conclusion to draw is in the form of a disjunction, an either–or. Either the theory is false, or there is something about the conditions of the testing that we don't understand. The observation, the measured result, underdetermines the decision whether it's the theory that's wrong or the conditions misunderstood. The rotation-curve data and the velocity-dispersion data underdetermine the choice between the general theory of relativity being false or there being dark matter.

Underdetermination is the norm when theory meets evidence. It's the root of uncertainty in science, planted directly in the logic of testing. It's a factor in detecting exoplanets, and it was there in the search for Neptune and Vulcan. But the case of dark matter is more radically underdetermined than any of these others. With Neptune and Vulcan the hypothesis of unseen mass was of objects that were at least observ*able* if not currently observed. And it was no unusual kind of matter, just one more of the same, one more planet. Planets exist; what's one more? But with dark matter, it's not just that this mass is unseen; it's that nothing like it has ever been seen, and never could be seen. It's not just the mass of the stuff but the existence of the stuff that is unobservable. With exoplanets, again, it's just a planet. There is little reason to challenge the theory of gravity used to interpret the data, since the existence of other planets is fully expected. Furthermore, theories on how planets are formed suggest that they should be relatively common. It would be a surprise if there *weren't* planets orbiting other stars, so there is good reason to adjust the mass distribution, rather than the theory of gravity, to fit the kinematic data.

What is dark matter? What is the stuff that makes up 27% of the mass in the universe and is hidden in galaxies and clusters of galaxies, causing the faster than predicted velocities? This is an open question. Some of it may be things we are used to and knew existed, just not in such quantity. Brown-dwarf stars and black holes in the clusters of galaxies would be invisible but gravitationally effective and could account for some of the disparity between predicted and observed rotation curves and velocity dispersion. Some, but not all. So there is also a hypothesized kind of dark matter that is not the normal stuff made of atoms, of protons, neutrons, and electrons. Protons and neutrons are in a category of elementary particles called baryons, so atoms, and things made of atoms, are baryonic matter. Some dark matter must be non-baryonic matter, made of elementary particles unlike protons and neutrons. These would be electrically neutral and immune to the strong nuclear force that binds protons to neutrons. They would interact only by the weak nuclear force, and, of course, gravity. The existence of these so-called weakly interacting massive particles is speculative, and it may require a new form of matter, not yet in the standard theories of elementary particles.

Recall the non-discovery of Vulcan. An unseen planet was suggested, maybe one always on the other side of the Sun and hence unobservable in principle. Diffuse matter was suggested, spread so thin as to be undetectable. These were the alternatives to challenging the theory of gravity, the Newtonian theory. And recall that Einstein described these as, "hypotheses which have little probability, and which were devised solely for this purpose," the purpose of reconciling the existing theory of gravity with contrary data. It's worth wondering what Einstein would make of the hypothesis of dark matter.

The flaws in the Newtonian theory were revealed when it was used in a new context. It had tested well on Earth and in the outer reaches of the Solar system, with Neptune, for example. But in the stronger gravitational conditions very close to the Sun, the orbit of Mercury, the theory failed. The general theory of relativity has been well tested by phenomena in the Solar system, but not at significantly larger scales such as galaxies or galaxy clusters. Thus, the accuracy of the theory in describing these larger phenomena is legitimately in question.

Testing a theory of gravity requires knowing the distribution of masses in the system. But, since dark matter can be detected only by its gravitational effects, the only way to know the distribution of masses, if some of it is dark matter, is by using a theory of gravity, presumably the one being tested. In fact, it's the Newtonian theory we have been using in the analysis of rotation curves and velocity dispersions. We could have used the general theory of relativity, and the results, the interpretation of the data, would be the same. It's what these two theories have in common that is being used in these cases. We need to identify what part of the theory is being used to locate dark matter. In at least one special case, the only aspect of the general theory of relativity that has this interpretive role is an aspect it has in common with *all* currently viable theories of gravity. The interpretive work is not done by the theory, *per se*, but by the more basic Principle of Equivalence.

We noted in Chapter 12 that the phenomenon of gravitational red shift is a consequence of the Principle of Equivalence alone. No other aspects of the general theory of relativity are involved. So, the measurement of the gravitational red shift tests only this one aspect of the theory. Similarly, using the red shift as an interpretive tool is using only the Principle of Equivalence, not the whole theory of relativity. If the red shift is used to interpret data, to detect dark matter, for example, this does not assume the truth of the general theory of relativity. It presumes only what is a more-or-less universal core principle, common to all contemporary theories of gravity.

Another common aspect of theories of gravity is the bending of light, gravitational lensing, and lensing is now an active way of detecting dark matter. Almost all theories, including Newtonian theory, predict some bending, but different theories predict different amounts. This means that using gravitational lensing to measure the amount of mass will require using the *right* theory of gravity, the one that includes the right amount of bending. Here again is the challenge, knowing that a theory is accurate in the conditions for which it is being used to interpret the evidence.

But it turns out that there is a way to use the phenomenon of gravitational lensing to detect dark matter that depends only on the *fact* of deflection but not the *amount*. Thus it is not the general theory of relativity, or any particular theory of gravity, that is doing the interpretive work. It is only a core concept that is common to almost all theories.

It doesn't have to be an Einstein's telescope that indicates *where* the lensing mass is. It could be a Newton's telescope, bending the light by the amount specified by Newton's theory, half that of the general theory. Determining *how much* mass is there requires a particular theory like the general theory of relativity. This means that lensing could be used to detect dark matter, independent of Einstein's theory, but only if the dark matter is separated from normal, luminous matter. In this case, the important question would be only about where the mass is, not the amount of mass, and virtually any theory will do.

In normal clusters of galaxies, the putative dark matter and the baryonic matter, that is, the stars and dust and gases, are mixed together. They have a common center of mass. But if two clusters collide, the baryonic matter will be separated from the dark matter. The intergalactic gases in each cluster will mingle and interact, causing some resistance and drag as they pass through each other. The intergalactic dark matter, on the other hand, will mingle but not interact. Dark matter is electrically neutral and so is immune to the electromagnetic interaction, the force that is largely responsible for the drag. The galaxies themselves are also electrically neutral and so far apart that they too pass through the collision largely uninhibited. They track with the dark matter while the gases get left a bit behind. In this way, the center of mass of the baryonic matter, most of which is in the intergalactic gases, separates from the center of mass of the dark matter as two clusters collide, pass through each other, and move apart.

An example of two clusters emerging from collision is the bullet cluster, so called because of the visual image it presents. It looks like a snapshot of the shock wave of a bullet passing through some viscous material (Schwarzschild, 2006, figure 2). The bullet cluster is at a distance 5 billion light years from Earth, and the two constituent clusters are separated by 2 million light years. The baryonic matter in the bullet cluster is detected by its electromagnetic radiation; it is luminous at various frequencies. The galaxies themselves are seen at optical wavelengths. The intergalactic gases are detected by their X-ray emissions.

The mass distribution of the bullet cluster can be located using gravitational lensing. It shows that the center of mass is significantly offset from the center of mass of the baryonic matter. In particular, the center of mass of each of the two clusters is ahead of the intergalactic gas. Since the intergalactic gases far out-weigh the galaxies themselves, the center of the gases is roughly the center of the baryonic mass distribution. But that's not where the lensing is centered. The mass needed to do the lensing is not where the bulk of the baryonic mass is, that is, not where the visible mass is. This is the key, and it provides the so-called direct evidence for dark matter. Some unseen matter must be doing the lensing.

Each of the colliding clusters acts as sieve for the other, separating dark matter, if there is any, from the normal baryonic matter, as an archaeologist's sieve separates artifacts, if there are any, from the dirt. Gravitational lensing then looks indiscriminately for stuff, to see where it is, whether it's in the sieve or in the dirt. In the case of the bullet cluster, there is mass in the sieve.

It is important to note that interpreting the data from gravitational lensing is very complicated in the case of the bullet cluster. The light that is bent by the mass of the cluster comes from objects that are not directly behind the cluster. They are off-axis. This results in images that are only slightly distorted, stretched out rather than rotationally symmetric like the Einstein ring we saw in Chapter 13. And the use of multiple sources of light requires a complex statistical analysis to determine even the simplest information about the lens. But, most importantly for the question of underdetermination, the analysis does not rely on the full theoretical content of the general theory of relativity. To locate the lens we need only that part of the theory that says that light is bent by mass, and nearly all viable theories of gravitation agree on this. The important point is that any lensing theory puts the center of gravity of the cluster – the centers, really, since there are two clusters separating – displaced from the baryonic matter. Any viable theory finds non-baryonic matter, dark matter, in the sieve.

A determination of how much dark matter there is in the bullet cluster would require more interpretive details of a full theory of gravity. To do that would require assuming the truth of the general theory of relativity, or some alternative, and would thereby preclude the possibility of challenging the theory. To use the metaphor of Einstein's telescope, we can tell where the lens is without using any particular theory of gravity, but we cannot tell the magnification of the lens without using the full theoretical force of the Einstein's theory or some other theory of gravity.

Theory is used to interpret scientific observations. That's fundamental to the method. The interpretive theory used should be both well-tested and independent of any hypothesis for which the observations are used as evidence. A careful look at using gravitational lensing to detect dark matter shows the importance of distinguishing among various components of a theory to see which are at work in the interpretive role. In the case of detecting dark matter in the bullet cluster, it is only the fact of lensing and not the amount that does the qualitative interpretive work. The rest of the general theory of relativity is idle.

There might still be some hesitation at calling the gravitational lensing *direct* evidence of dark matter. It is after all observing something other than the dark matter itself. It requires an inference to the existence of the dark matter from the way it alters the image. In what is called indirect evidence, the dark matter alters the motion of the things we observe, the galaxies in a cluster, say. In this new direct evidence, the dark matter alters the light from the object we observe by bending it in a gravitational lens. Which is the more direct? It shouldn't matter. This physical difference is not important to the quality of the evidence or the value of the test. What is important is the status of the theoretical account of the interaction that delivers the information. The theories used should be independent of claims about dark matter and should be independently justified. The Equivalence Principle and the prediction of lensing pass both tests, and this is what gives the claim of detecting dark matter credibility in the case of the bullet cluster.

There is still uncertainty, and still the disjunction. Either there is dark matter in the bullet cluster or light is not bent by a gravitational field. So this is neither a confirmation of the existence of dark matter nor proof of gravitational lensing. But it shows the consistency of the two concepts, and the gravitational lensing brings independent justification into the link.

15

The Structure of Scientific Knowledge

A good scientific theory will show how some of the pieces of nature fit together. A good science class will do the same. Teachers use tools like comprehensive exams or term papers to achieve that kind of coherence and overview. With that as a model, this final chapter will be in the style of a term paper, the kind that requires both summary and reflection on the most important ideas developed during the course, or in this case, the book. The idea is to make sure that no important component of scientific method is overlooked, and to see how the pieces fit together. Don't panic, and don't despair; this will be both painless and valuable. We'll restrict the topic to scientific method, using the scientific accomplishments about gravity only to clarify the methodological conclusions.

Some term papers are really good. One exciting and humbling aspect of teaching is working with lots of people more talented than yourself. And you get a new group each year. Some of the best writing and reasoning I have read have been in the papers by students. That's the style of term paper we're going for here.

A term paper is only as good as the assignment, the clarification of the topic, and the expectations. If the directions are clear and the task is focused and reasonable, the project won't be too difficult and the results will be both readable and informative, something of which to be proud. So, here is the assignment.

At this point you should understand the fundamentals of the science of gravity. You should also be familiar with the most important aspects of its history, the development of major ideas and the changes from one theory to another, and the ways in which these theories were, or were not, tested. Use these scientific and historical details to clarify and defend a description of scientific method as it applies to the science of gravity.

It is unacceptable to present a method based on vague slogans or undefined terms. Don't rely on clichés. Anyone will admit that scientific results must be based on evidence, but what do you mean by "based on," and, for that matter, what do you mean by "evidence?" And it's easy to say that science requires hard facts, but what makes a fact hard, and is there any role for soft facts? Your account of scientific method should clearly derive from, and apply to, the science of gravity, so refer to features of that science to make the method clear.

Extra credit: Does the method of the science of gravity apply to other sciences such as biology and geology? Is it a general scientific method?

Optional alternative: You have the option to argue that there is in fact no single method that covers the science of gravity, and hence no such thing as a general scientific method. If this is your thesis, make sure it is not motivated simply by laziness or a vague cynicism about science or conformity. Don't be glib; you need a cogent argument and evidence. A conclusion that there is no method must be supported by details from our study of gravity. (I am not blind to the irony in this requirement. It insists on the same method to justify the conclusion that there is no method as to justify the description of the method.)

And now, the term paper.

I will argue that there is a clearly identifiable method in the science of gravity, and I will do the extra credit. I will argue that all aspects of the method implicitly refer to making multiple connections in a network of theoretical and evidential claims about nature, and that this multiplicity and variety of connections tips the balance of probability that the components of the network are true. Scientific method is all about the far-flung, tangled web.

But first, consider the optional alternative, the case that there is no identifiable scientific method. This may have a tacit assumption that the burden of proof weighs on the claim that there is a method, so no argument is needed. Maybe just a snappy quote will do the trick, one with authority and blunt dismissal. It is common among non-methodists to cite Percy Bridgman, a Nobel Prize winning physicist and contemporary of Einstein. "The scientific method, as far as it is a method, is doing one's damnedest with one's mind, no holds barred" (Bridgman, 1955, p. 535). The "no holds barred" is misleading in its suggestion that there is neither structure nor standards in scientific reasoning. Doing what with one's mind? Bending spoons? Numerology? Simply meditating? Bridgman had more to say about method, and the quote by itself is a bit of an insult to both the man and the institution. It's vague and dismissive, and if you work to make the vague concepts more precise you will find that exactly the details of a scientific method emerge. These slogans that make it sound as if there is no special method to doing science are worth going over. They are a valuable foil, and correcting or simply clarifying them will be a good way to start the description of what the method looks like.

Another popular authority is Einstein. "The whole of science is nothing more than a refinement of everyday thinking" (Einstein, 1936, p. 349). The refinement is the crux. And surely not every sort of everyday thinking shows up in science. Wishful thinking, something we do every day, even if refined is not part of scientific method. Scientists, since they are people, certainly bring wishful thinking into the process, but it's the method, the logic of testing, controlled experiment, peer review, and so on, that is meant to minimize this cognitive vice. The same is true for superstition, denial, jumping to conclusions, cherry-picking and other weaknesses of the mind. Scientific method tries, with more or less success, to exclude these examples of everyday thinking and to refine the others.

The Bridgman and Einstein references have one thing worth emphasizing. They both say that what is common among the sciences is about what scientists think, not what they do. The common method is not to be found in the physical objects or activities of the different sciences, but in the reasoning and conclusions associated with the activities. Astronomers don't use Petri dishes; they don't really do controlled experiments of any kind. But something about their reasons for doing what they do is the same as those of a microbiologist. The common method is not in the similarity of actions but in the structure of knowledge. That's what we're looking for.

It's worth considering more substantive arguments that there is no single scientific method. Note, for example, the important differences among the various theories and theorists of gravity and ask if these don't require different methods. Einstein explicitly pointed out a distinction between a theory of principle and a constructive theory, and the difference looks to be methodological, noting how the theory is developed and the theorist's attitude toward evidence. Aristotle and Einstein gave us the examples of theories of principle, while Newton supplied a constructive theory. But once each of these theories was on the books the scientific expectation was the same, namely comparison between theoretical predictions and new evidence. Neither Aristotle nor Einstein as individuals did this, but scientific method requires it no less for the principled than for the constructive theories. Aristotle and Aristotelians are now faulted for this lapse because it is a failure of the general standards of scientific method. Einstein's theory was tested by other members of his scientific community, and this makes the group project scientific. Some manner of empirical testing is a necessary component of any scientific theory.

The contrast between a theory of principle and a constructive theory is in the derivation of the theory, the source of the ideas. It's in the context of discovery. There probably is no single method for coming up with a good scientific idea, and this is where creativity, personal quirks, luck, and doing one's damnedest are helpful and unregulated. The methodical part of science comes next, in the context of justification. This is where the logic is enforced, contradictions must be eliminated, evidence is required, and overall coherence is the standard. It's in the context of justification where we've found the general method.

When we find what all the sciences do have in common it will be in the connections, between theory and observation and among various theories. It will be in the structure of how the descriptive claims are held together and the standards of fitting into what Weinberg called the tangled web. It may be more accurate to call the result the structure of scientific knowledge rather than the scientific method. But it describes what's going on in the mind when scientists are doing science; it elaborates the glib remarks of Bridgman and Einstein, and in that sense it's about the method.

What do the different contributions to the science of gravity have in common? These will be the basic components in the structure of knowledge and the start of the description of scientific method. There will be no procedural flow-chart, no step-one, no recipe for scientific success. It's not that kind of method, and the presentation will be nothing like the science-textbook caricatures.

All contributions to the science of gravity involve theoretical and observational ideas in mutual, reciprocal support. The science is not restricted to observables, contrary to some scientists' assessments. For example, "As in any other scientific study, we are only able to describe what we can observe, . . . " (Schutz, 2003, p. 346, ironically just pages after mentioning dark matter). Nor is any of the science pure theory. There are always both kinds of information, describing both what can and cannot be observed.

Aristotle cites direct observation in claiming that unsupported objects fall straight down and that a heavy one falls somewhat faster than a light one. He links this to the theoretical claims that the unsupported object has a natural motion and a goal, both directed toward the center of the universe. None of that is directly observed. The observational information of straight free-fall motivates and confirms the theoretical concepts of natural motion (the free in free-fall means natural) and the center of the Earth, coincident with the center of the universe, and the directing goal. The theoretical information influences the interpretation of what is observed. The heavier object is expected to hit the ground first, and so that is a relevant detail in the experiment. Theory and observation influence each other.

Newton had a wealth of observational information to tie into the web, including Kepler's laws and Galileo's time-squared law. These were the empirical antecedents of the theory of universal gravitation. Subsequent observations of planetary and projectile motion were also in the structure. The theoretical components were most notably the concept of the force, in itself invisible, and the instantaneous action at a distance. The observations certainly support the theory, both in Newton's derivation of the formula for the force of gravity and then its testing. And there is the characteristic influence of theory on the observations. Pursuing the analogy between a falling apple and the orbiting Moon would be a fanciful waste of time unless there was some good reason to think that gravity was universal and linked to nothing other than the mass, not the composition, of the objects. Newton knew what to look for and what mattered, and he used this theoretical knowledge to both design and interpret the observations.

Like the others, Einstein worked with some basic observations. Heavy things and light things, in appropriate circumstances, fall to the ground at the same rate. Experiments in a freely falling elevator are experiments with no gravitational effects. The precession of the orbit of Mercury is 38″ per century more than expected. And so on. And of course, he added theory to the mix, including unobservable components like geodesics in four-dimensional spacetime. The observations influence and give credibility to the theory, as the theory is used to interpret the evidence. Calling an event of gravitational lensing "Einstein's Telescope" makes clear the interpretive role of the general theory of relativity in using data from lensing.

Theory and observation are the bricks in the structure; we need to understand the mortar and the architecture. And we need to include a third component, a third kind of information that needs to be accommodated in any coherent system of scientific knowledge. It's what we have called conceptual criteria for evaluating scientific ideas. These are sometimes aesthetic, as in the beauty of a scientific theory, or pragmatic, as in simplicity, or metaphysical, that is, abiding by some basic metaphysical prescription like perfection in heaven. Aristotelians required perfect circles in any astronomical model. Galileo cited the greater complexity of the Ptolemaic model – moving all the stars and planets when you could get the same effect by rotating just one object, the Earth – as one reason to reject the old world system. Newton was motivated by a presumption of uniformity in the universe, that a universal description of things was more likely to be true than a patchwork. And Einstein often seemed more impressed by the beauty and principles at the core of relativity than by any empirical success. These sorts of conceptual influences can't be ignored in describing scientific method.

None of the three kinds of information, the empirical, theoretical, and conceptual, is foundational in the sense of having unquestionable authority over the others. No single scientific claim is immune to revision. The structure of scientific knowledge does not rest on solid ground; it is the tangled web and it gets its strength, the reason to believe the whole and any of its parts, by being tightly woven. The network of ideas, whether it's Aristotelian, Newtonian, or relativistic gravity, must be coherent.

A coherent system of beliefs and a coherent theory would be, at a minimum, consistent. Contradiction is nonsense and it demands revision or removal of inconsistent claims. Of course, it doesn't indicate which of any contradictory claims must go, and a new idea that is inconsistent with an established network may warrant revising or removing the network and rebuilding around the new. Consistency is a strict rule, but its enforcement will always be ambiguous in this way and hence it will leave room for some discretion. Newton's and Kepler's mechanism of action at a distance was inconsistent with established theories of action by contact, yet the Newtonian theory of universal gravitation allowed it into the description of nature.

There is a more robust measure of coherence, one that is more than just a necessary aspect of scientific knowledge but a contributing reason to believe things are true. Ideas should be not only consistent but also actively connected by logical and explanatory implications. Irrelevant and independent ideas can be consistent, but one idea implying or being implied by others shows the entanglement in the web. The more such connections and the fewer loose ends, details that can be changed without consequence, the more we can claim to understand what's going on in nature. The tangled web shows how things not only do fit together but must fit together. This is making sense. Remember that our first act of doing science together was relating atmospheric pressure to wind velocity in the discussion of fields. It was showing how these fit together. This is coherence.

Coherence and making sense are relative to a person's and a community's existing system of beliefs. They are relative to what Kuhn, and now just about everybody else, calls a paradigm. What makes sense in the context of Aristotelian science is what fits into the network of Aristotelian ideas. The same is true for any other coherent system of ideas. So again, making sense may be necessary but it can't be sufficient as a scientific standard, since multiple possible paradigms allow multiple, potentially incompatible, belief systems, each of which makes internal sense.

So, there must be more. It's in the details of the implications and explanations, the nature of the links between theory, observation, and conceptual constraints. Theory has to make sense in light of the evidence, and evidence has to make sense in the context of theory. In the connection between theory and observation, neither is the foundation for the other; their support is mutual. Observation supports theory in three possible ways: testing the theory, using the theory to explain phenomena, or just plain using the theory to get things done. The logic is the same in all three. That is, the form of the link between theory and observation is the same. It is always an if–then entailment and the theoretical claim is always the if. If the theory is true, then this is what will be observed. When it is observed, a link in the web is made. More links are better, and a greater variety of observed phenomena is better still, as these structural features show tighter and more far-reaching coherence. On the presumption that nature is systematic and interconnected, the coherence is an indication of the likelihood of a match between the descriptive network of ideas and the way things really are. You can never be sure, and this lingering uncertainty is in the logic itself.

Everyone knows that you can't prove that a theory is true, although not many know why. Here is a misleading explanation of scientific uncertainty that misses the important point. I heard this on the first day of a biology class, a quick argument that a theory cannot be proven but it can be disproven. Consider a scientific theory of the logical form: All As are B. Test this by looking at a bunch of As and seeing if they are B. If you find an A that is B, that's good, but it surely doesn't prove that *all* As are B. Even if you find lots of As and they're all B, you still could have overlooked that one A that isn't a B. It's the "all" in the law that makes it unprovable with finite evidence. The theory cannot be confirmed, if that means proven beyond all doubt. But it can be disproven, falsified, by finding just one A that is not B. Again, it's the "all" in the law that makes it vulnerable.

What's wrong with this picture? It assumes that both the As and the property B are observable. The data of an A being B or not B must be unambiguously observable and beyond challenge of interpretation or mistake. But this is not characteristic of science. Theories are about unobservable As and B, or at best indirectly observable As and B. The science of gravity, for example, has theories about fields, space-time, and curvature. The observations are not directly of what the theory describes. You can't observe A to see if it's B or not B.

The real reason that it is impossible to prove a theory is in the logical relation between theory and evidence, the if–then relation. To conclude that a theory is true on the basis of a true prediction is the fallacy of affirming the consequent. That's a formal way of saying that there are always alternative theories that make the same true prediction. But the reason that *dis*proving a theory is just as impossible is quite different. It's in the necessary complexity of both the theoretical implication and the evidence. In real cases, like the tower argument, Tycho's search for stellar parallax, or the discovery of Neptune, the hypothesis makes an observable prediction only with the informational assistance of some other theoretical or conceptual ideas. Not finding what you're looking for implicates these other ideas no less than it does the hypothesis, so you can spare the hypothesis by blaming an assistant. Or you can doubt the evidence itself. Observations in the science of gravity are generally indirect and require interpretation. Maybe that's not really an A, or it's a variant form of B.

Observations in science are interpreted under the influence of theory. It takes expertise to know what the data mean, and expertise requires being fluent in current theory. The examples from the science of gravity show a variety of ways in which evidence bears the influence of theory. Scientific observation is neither complete nor random. Some understanding of how nature works directs the selection of what to look at and what to look for. The falling apple was noteworthy to Newton as to no one else, since he was prepared to find a dynamic connection to the Moon. Aristotle recorded the important fact that the heavier object hits the ground before the lighter. These were good data, consistent with theoretical expectation. And Galileo used a ramp to observe properties of free-fall, apparently understanding that he could safely ignore the differences between the natural and constrained motions. In all these cases, some theoretical authority pointed out the importance and relevance of the observations.

Theory also directs the controls in a controlled experiment. Experts, the theoretically savvy, know which properties need to be controlled and which do not.

Galileo's ramp again. Pay careful attention to the angle of the incline but not to the composition of the rolling ball or even its size. Applying exacting mathematics to a messy world demands focusing on some quantities while ignoring others.

Properly controlled experiments and observations made under the right conditions provide reliable data, and theory rules on propriety. Did Galileo's telescope produce accurate images of objects in the sky? That depended on which worldview, which web of theories, one held. The interpretation was not so much in the eye of the beholder as in the mind of the beholder. The solar-eclipse data in 1919, the first test of the general theory of relativity, were similarly vulnerable to interpretation. Some of the results were ruled out as unreliable, and the ruling was made with the theory in mind. This is not bad science; it's normal science. Good science requires good data, and it's expert scientists who can tell good from bad. They are experts because they are trained, and an important component of the training is knowing the current theory.

Galileo apparently saw Neptune in 1612 and even drew the image into his star chart, but he is not credited with discovering the planet. He didn't see it as a planet; he thought it was a star. Scientific observation is not just the physical perception of something; it is also the description of what it is. To observe Neptune is to observe it as a planet. You have to know what the basic data mean. And this step, from just seeing it to seeing it for what it is, requires a theoretical context. It requires interpretation. This is another important way in which theoretical ideas in the tangled web influence observational ideas. Neptune was seen as a planet when it was expected as a planet.

All of these ways in which theory is, and must be, used to interpret are characteristic of the informational connection from theory to observation. There are also connections from one theory to another. Two theories in a system must be consistent. Galileo revealed an internal inconsistency in the Aristotelian theory that a heavy object falls faster than a light one. It ended up with a composite object falling both faster and slower than one of its components. But then Galileo allowed an inconsistency between his own theory of the tides and the Principle of Relativity. The inconsistency shows that some part of the theoretical network must be wrong, whether it's Aristotle on natural motion or Galileo on tides. This necessary inter-theoretical link was missing in both cases.

Beyond consistency there is entailment among theoretical claims. Theoretical accounts of electricity and magnetism together with the Principle of Relativity entail that the speed of light is absolute. And from there, a spatial property like at-the-same-place being relative entails that the corresponding temporal property, in this case simultaneity, is relative. With so many of these kinds of theoretical connections, relativity is, again citing Weinberg, theoretically rigid. There are very few flexible parameters; almost no detail can be changed without forcing far-reaching changes in the network of ideas. Contrast this with the Ptolemaic system in which almost all properties of the important components, the epicycles and deferents, could be individually adjusted without affecting any other aspects of the model.

The more tightly coherent a theoretical system, the less empirical input it requires. Properties are determined in principle. Valuing this inflexible structure in scientific knowledge assumes that nature itself is systematic, and that properties are inter-related. One property determines others (in the factual sense of determines) and so when we can determine (in the knowing sense) one property from others it is likely that we are seeing the way nature is put together. It's not just the way things happen to be; it's the way they have to be.

The structure of scientific knowledge is meant to reflect the structure of nature. There is a variety of ways that theoretical and observational ideas are connected and related, different ways to entangle the web. Again under the assumption that nature itself is not a loose, disconnected, random collection of events, the more connections we can secure in the system of descriptive claims, in the science, the more confident we can be that the knowledge is likely to be accurate. Coherence is indicative of truth, and the scientific method, its goal and the standards to evaluate its application, is to maximize the coherence.

There is one more kind of link in the network, one that is often overlooked or undervalued. We recognize links like testing, explanation, and application that connect theory to observation. Less apparent, but hard at work in the science of gravity, is the use of analogy. There are some explicit cases. Galileo deployed analogies in the tower argument, first saying that the moving Earth is like a moving ship, and then that the stone on the ship is like a marble rolling on a mirror. Newton developed the analogy between a falling apple and the Moon. In the theory of relativity we talk about gravitational lensing, taking advantage of the comparison to optics. And modern textbooks often describe curved spacetime as being similar to a trampoline with a bowling ball in the center, stretching the surface in a way that draws other objects in. The use of analogy is so pervasive and varied in science that understanding scientific method requires understanding scientific analogies.

Every argument by analogy follows the same logical form and it is different from the logic of testing and explanation. The core idea is that one thing is like another, so what we know about the one is likely true of the other. The inference is manifestly uncertain, and there is always an implied "probably" in the conclusion. The one thing is *like* the other, not *identical* to the other. They don't have all features in common, and there's the risk that the conclusion is about one of the ways in which they differ. An argument by analogy is a way of extending knowledge, from what we know about the one thing to what we don't know about the other. This will always be risky to one degree or another. The challenge is to evaluate and minimize the risk.

Of the two things that are alike, the Earth and the ship, the apple and the Moon, spacetime and a trampoline, one is relatively well understood and familiar while the other is not. The contrast may be because the one has been observed while the other has not, and this may be because it is in principle unobservable, or because one is tightly embedded in a theoretical structure while the other is a new idea, a hypothesis. A falling apple is easy to see and measure, and the experiment can be repeated and controlled, but the orbiting Moon is distant and indifferent to any experimental manipulations. The apple is familiar, and the argument is to extend what we know about it to the unfamiliar, the Moon. That's the essence of argument by analogy. It's hard to tell whether the Earth is moving or not, because we are stuck on the thing itself, or we were at the times of Aristotle and Galileo. It's easy to tell whether a ship is moving or not, because we can get off and stand on shore to look back. This gives Galileo empirical information about the ship that the analogy can extend to the Earth.

The form of every argument by analogy is the same, and this gives it a standardized role in the scientific method, but there are multiple uses of these arguments. Textbooks use analogies to clarify ideas but not to prove or make them credible. Newton used the apple analogy to propose the inverse-square dependence and universality of gravity, but not to close the deal and prove the law. Galileo used the tower argument to prove a point, not that the Earth moves but that the trajectory of a falling stone could not demonstrate, one way or the other, whether the Earth moves.

It's important to note the various roles of analogy in the scientific method, because analogy is often mistakenly described as doing only minimal and unimportant work. Perhaps because of the explicit uncertainty in the connection between known and unknown, analogy is sometimes dismissed as merely explanatory, that is, for the benefit of non-scientists. That's what the trampoline is doing in the middle of the general theory of relativity. It doesn't advance the science; it only brings it into the living room. This would put analogy in neither context of discovery nor justification but in a third context, call it the context of public education. It's not that this is not worth doing, but it's not within the business of science. We have plenty of examples, though, that show analogies hard at work in the scientific process. It is not merely explanatory.

Analogy is used in science to provide reasons to believe a theory. The discovery of Neptune was a real success and it became an exemplar for interpretation of astronomical evidence. The prediction of gravitational theory was off in that case, and the problem was not in the theory but in some missing mass. Apply this reasoning to similar cases to give some good reason to think there is missing mass in those cases as well. Mercury was like Uranus in that in both cases the orbits did not exactly follow the Newtonian predictions. This was reason to hypothesize missing mass in the Mercurial case. This justified the search for Vulcan. Galactic rotation curves are like Uranus, again because they defy Newtonian and relativistic predictions. So again, based on the similarity, the analogy, there is reason to believe there is missing mass. In this case it's dark matter. The important point is that analogy is doing influential work in interpreting the data and supporting the reasonable hypothesis. It's at work in the context of discovery.

To say that gravity is a universal phenomenon, a force or a metric variation, is to invoke an analogy. It extends what we know about things we have studied, apples, stones, the Moon, planets, some galaxies, and so on, to things we have not studied. The "universal" is an implied analogy. More specifically, stars that are suspected of holding extra-solar planets are regarded as being similar to the Sun. The gravitational effects here will apply there, based on the analogy. The star wobbles for the same reason the Sun wobbles. This is a very plausible extension of local knowledge to a distant and global scale, and the plausibility is based on the reasonable analogy. This shows that analogy has a very pervasive role in scientific method.

Analogy is everywhere in science. Every controlled experiment is built around an argument by analogy. This should count towards the extra credit, since controlled experiments are the business of most sciences. The experiment is meant to reveal some small aspect of what happens in nature, and that means in uncontrolled nature, when we're not looking and not holding any properties steady. This is possible only if there is good reason to believe the uncontrolled situation is relevantly like the controlled conditions in the lab. That's an analogy, and the careful argument includes the reasons that the specific controlled properties, Galileo's inclined ramp, maintain the relevant similarities to the uncontrolled, free-fall. These reasons must be supported by some theoretical understanding of both sides of the analogy; they can't be purely empirical, since the uncontrolled side of things has not, and maybe cannot be, observed. So, this is another way in which theory is used to interpret evidence. The differences between the controlled and uncontrolled cases don't matter to particular results being studied. The evidence can be applied from one case to conclusions about the other.

All of this is meant to show that analogy is an important component of scientific method, in both the links from theory to observation and from one theory to others. It's the basis for extending knowledge from one context, the laboratory or the Earth, say, to others, to the natural world and extraterrestrial phenomena. It is the essential tool in making claims that are general and universal, that is, in making laws. And it is explicitly uncertain.

Uncertainty has been the rule in science, at least since Galileo. The method is imperfect and no conclusions are above doubt. But that doesn't mean they should all be doubted. There is a lot of room for good reason between certainty and pure guesswork. That's exactly where the scientific method operates, to locate a claim on the spectrum of justification.

Avoiding that fallacy, the one that goes, if you can't be sure then you have no good reason at all, is helped by clearly marking the difference between a belief being true and its being justified. The concept of being-true, or being-false, is easy to clarify, despite whatever smoke and mirrors you might have encountered in a philosophy class. How many planets are there in our Solar system? There is a right answer to this, and that's all we mean by true. It requires that the terms are precisely defined, clarity in particular on what counts as a planet, but then the answer is true or false depending on whether it matches the actual arrangement in nature. There is a planet beyond Uranus; that's true. There is a planet closer to the Sun than Mercury; that's false. The same standard of true-or-false applies to more complicated claims. Why does a stone fall to the ground? There is a right answer to this as well.

True-or-false is not a matter of degree. A descriptive claim, theory or observation, is either true or it's false. Justification, the good reason to believe it's true, does come in degrees. There lies uncertainty. One successful prediction does not prove a theory, but it does add a little justification. If the patient has a fever, that's *some* reason to think she has the flu. Additional successful predictions, of a variety of phenomena, add more justification. There will never be complete justification in the sense of there being no chance the theory is false. Furthermore, it is possible for there to be a true theory that has no justification. A drunken Greek at the time of Aristotle who claimed that the Earth is spinning would have been absolutely right but without justification. It's also possible to have a well-justified theory that is false. Newtonian universal gravitation is a good example. There was very good reason to believe there is a force between massive objects, but this is false, at least according to the general theory of relativity.

The distinction between truth and justification is related to the two senses of determining a property and, not surprisingly, to the hard-working concept of under-determination. To say that a property like the acceleration of a stone or the number of planets in the Solar system is determined to be x could mean either of two things. It could mean that there is a value for x, a fact of the matter in nature. Or it could mean, in addition, that we know the value of x. A relative property is not determined in either sense without a specific reference frame. Not only can we not know the velocity of the stone without a reference frame, it simply doesn't have a velocity independent of a reference frame. An absolute property is always determined in the first sense, the physical sense, but not determined in the second, the knowing sense, unless we make the measurement. The physical sense of determine is about nature, independent of us. It's about truth. There is a value of x, and a true statement has the right value of x. The knowing sense of determine includes us. It's about justification. It's about our getting the information on the value of x.

Underdetermination, as we have been using the term, is about the knowing sense of determining. On questions about gravity, there is a correct theory. Nature determines the facts, but we never have enough information to fully determine what the facts are.

There's a lot to do in trying to determine (in the knowing sense) the truth about nature. There are all the different kinds of connections between theoretical and observational claims to secure, and there is an obligation to continue and find more connections. It takes a lot of people, and no individual scientist is expected to work on all aspects of the method. Einstein never tested the general theory of relativity, but the theory has been tested. You never see any one scientist doing all that it takes to be scientific, and that may be why some scientists deny that there is a single scientific method. The structure of scientific knowledge is, and must be, holistic and global. Consequently, the method is a group effort.

And this leads to the extra credit, extending the description of scientific method, or the structure of scientific knowledge, beyond the specific case of gravity. Two brief (this is extra credit) arguments will show that there is a general method shared by all the natural sciences.

All of the conclusions about the structure of scientific knowledge that have been proposed have been based on principle. It's not just that all observations in the science of gravity happen to be interpreted under the influence of theory; they *have* to be. The requirement, and hence this link between theory and observation, is a result of the need to select, certify, and give meaning to data if they are to contribute to knowledge. And the logic of testing *has* to be of the if–then form we found in the case of testing theories of gravity, with the theory being tested in the if-position. Furthermore, extending knowledge from one context to another, a fundamental and universal goal of science, *has* to involve argument by analogy. All of these aspects of scientific method were brought to light in the examples of gravity, but the examples served as motivating and corroborating what are fundamentally logical principles. So, our study of the science of gravity results in a methodological theory of principle, and the principles apply to any inquiry of nature that claims to draw conclusions beyond what is immediately observed. The results apply to any natural science.

The second argument is really the same as the first, just in a different form. It's an argument by analogy. Biology is in many ways like physics. Both study the natural, physical world. Both look for and present generalizations about nature, some that are about objects and events that cannot be observed. Both aspire to form and retain theories that are true, and they are careful to retain those theories for which there is good reason to think the theory is true. They accept theories on the basic of justification. So, biology and physics have the same challenges, goals, and informational resources. One of them, physics, we have seen, has a method for meeting the challenge, so by analogy biology (and geology, chemistry, and other natural sciences) probably use very much the same method.

The structure of scientific knowledge and the method for putting the structure together is, as Einstein suggested, very much like what we all do on a daily basis when we are trying to know about our physical surroundings. We observe and theorize and test and explain, and we reason by analogy. We do other things as well, like guess and stereotype and jump to conclusions. Scientific method is directed to nurture the good reasoning and weed out the bad. It's common sense but done more carefully and rigorously, and with a commitment to abide by reason and evidence. It may be just a "refinement of everyday thinking," but the refinement is important.

Being scientific, that is, using scientific method, means taking full advantage of all kinds of connections among theories and observations. Evidence is *scientific* evidence insofar as it is collected, certified, and interpreted with all things considered, that is, under the influence of relevant background knowledge. Unscientific evidence is naïve, in the sense of being ignorant of theoretical context. It stands alone, and that isolation is the fundamental lapse in method. Scientific method is the cooperative group doing their damnedest to connect all the pieces, empirical and theoretical, no holds barred.

Glossary

acceleration of gravity (*g*)   All objects in free-fall near the surface of the Earth accelerate downward at the same rate, independent of their mass or composition. This rate is the acceleration of gravity, *g*.

$$g = 9.8\,\text{m/s}^2$$

action at a distance   A causal influence between one thing and another that has an effect even when the objects are some distance apart and with nothing in between is said to be an action at a distance. It is generally assumed that the cause and effect are simultaneous, so the influence moves with infinite speed. Newtonian universal gravitation is an example of a force that requires action at a distance.

affirming the consequent   This is the name of a logical fallacy, a particular form of argument that is always invalid. Being invalid means that it is possible for all the premises of the argument to be true, yet the conclusion is false. The premises, in other words, do not prove the conclusion. The specific form of the argument that is the fallacy of affirming the consequent is as follows.

premise	If P then Q
premise	Q is true
conclusion	P is true

analogy, argument by analogy   An analogy is a claim that one thing is like another. An argument by analogy uses the similarity between two things to draw a conclusion about one from what is already known about the other. The basic form of an argument by analogy is as follows.

premise	x is like y
premise	y has property P
conclusion	x probably has property P

An argument by analogy is inherently uncertain.

centripetal acceleration   This is the acceleration of an object that is turning or in orbit. It is the change in the direction of the object's velocity, not the change in the magnitude of the velocity, the speed. Centripetal acceleration is always directed toward the center of the turn or the orbit.

centripetal force   This is the force that produces centripetal acceleration. The term "centripetal" does not describe the particular kind of force, as in gravitational or electrostatic or tension in a string. Rather, it describes how the force, regardless of source, is used. It is used to cause the object to turn inward, toward the center.

coherence   The coherence of a system of ideas and statements is a measure of how well connected the components are to each other. The connections include logical relations such as consistency and implication (one statement implies and is implied by others), and relations between theories and evidence such as testing and explanation. Maximizing coherence in a scientific description of nature is the basic goal of scientific method.

context of discovery   This is a philosophy of science term, roughly describing that part of the scientific process in which a new idea is first proposed. In this context there is no clear method, and different scientists have used, or stumbled upon, very different ways to come up with new ideas.

context of justification   This is a philosophy of science term, describing that part of the scientific process in which ideas are tested, revised, rejected, or accepted. This is where science is methodical and general standards on the acceptability of theories are applied.

Copernican Principle   Neither human beings nor our position in the universe is unique or special in any way. Nature is largely the same here as elsewhere. The universe is, on the large scale, homogenous. All of these are ways of stating the Copernican Principle.

covariant   A relation of multiple properties that has the same form in every reference frame is covariant. The values of the individual properties may be different in different reference frames, but together they co-vary, that is, they vary in a correlated way such that their overall form stays the same.

dark matter   Matter in the universe that has mass, and hence interacts gravitationally, but that is otherwise undetectable is dark matter. Different theories propose different kinds of dark matter, from the exotic, such as entirely new kinds of particles, to the mundane, such as brown-dwarf stars.

deferent   In early models of planetary orbit, a planet orbited on a small circle, the epicycle, the center of which orbited on a larger circle, the deferent. The radius of the deferent determines the average distance between the planet and its orbital center.

Doppler shift, Doppler effect   Any signal that travels as a wave will have a longer wavelength when the source and receiver are moving away from each other, and a shorter wavelength when they are moving toward each other. This is the Doppler shift. Two noteworthy examples are the change in pitch of a siren as the vehicle changes from coming toward you to going away, and the so-called red shift in the frequency of light from distant galaxies as a result of their moving away from us in an expanding universe.

dynamics   The systematic study of the causes of motion is dynamics. Dynamic properties are things like force and mass. Dynamics together with kinematics, the precise description of motion, form the branch of physics called mechanics.

eccentric   In early models of planetary orbit, the center of the orbit was off-set from the Earth. The point of the orbital center, and the distance from the Earth to this point, are both referred to as the eccentric.

epicycle   In early models of planetary orbit, the planet orbited on a small circle, the epicycle, the center of which orbited on a larger circle, the deferent. If the radius of the epicycle is large enough, and the period of orbit on the epicycle is short enough, the path of the planet is a rosette.

equant   In the Ptolemaic model of planetary orbit, the angular speed of orbit is a constant, but not around either the Earth or the geometric center of the orbit. It's constant around a point referred to as the equant.

escape velocity   This is the minimum speed it would take to break entirely away from the gravitational attraction of some massive object like a planet or a star. Any slower and one would be held in orbit or eventually fall to the surface.

extra-solar planet (exoplanet)   Any planet in orbit around a star other than the Sun is an extra-solar planet.

field   A field is a continuous distribution of values of a particular property at different points in space and at different times. In other words, it is a property as a function of position and time. In physics, a field can play a role as if it has physical reality. Waves in the electromagnetic field, for example, carry energy and mediate interactions between electrically charged particles.

geodesic   The line that is the shortest distance between two points is the geodesic. On a flat surface, a geodesic is a straight line. On a curved surface, the geodesic itself will be curved. A taut string on the surface of a sphere follows the geodesic between its two endpoints.

gravitational lensing   Gravity bends light. This basic effect is called gravitational lensing. When the gravitational field is very strong, light can be bent enough to produce magnified or distorted images of the source. Both the Newtonian and relativistic theories of gravity include this effect, but by different amounts.

gravitational radiation   In the general theory of relativity, some kinds of movement of a gravitational source will create waves in the metric field that propagate out, carrying energy. This is gravitational radiation. It is not a feature of the Newtonian theory of gravity.

gravitational red shift   Gravity affects time duration. The wavelength of electromagnetic radiation gets longer as the wave moves out from a massive object. Visible light will be shifted toward the red end of the spectrum. Clocks and all period phenomena will tick more slowly closer to the gravitational source. All of these effects are the gravitational red shift. Both the Newtonian and relativistic theories include this effect, and to the same amount.

inertia   An object will naturally resist any change in its straight-line, uniform motion. This is inertia, and the more mass the object has, the more inertia.

inertial reference frame   A reference frame that is moving at a constant velocity is inertial. In the general theory of relativity, an inertial reference frame is any frame that is in free-fall in the local gravitational field. In an inertial reference frame, geodesics are straight, thus the spacetime is flat.

instrumentalism   Instrumentalism is an attitude about the value and limitations of science. It claims that theories are just tools (instruments), ways of thinking about what we cannot observe, that allow us to make sense of what we can observe. They are valued for being useful, not for being true.

intrinsic curvature   A surface or higher-dimensional space has intrinsic curvature if the formula for determining the distance between two points changes at different locations. In other words, the distance formula, the metric, is itself a function of location. Intrinsic curvature is determined without reference to the surface or space being embedded in an even higher-dimensional space.

invariant   A property with the same value in every reference frame is an invariant. For example, the speed of light is invariant.

inverse problem   Trying to determine the nature and magnitude of a cause by using the observed details of the effect is an inverse problem.

kinematics   The precise description of motion is called kinematics. Kinematic properties are things like position, velocity, and acceleration. Kinematics together with dynamics, the systematic study of the causes of motion, form the branch of physics called mechanics.

Mach's Principle   All spatial and temporal properties of things, for example, position and velocity, are determined in relation to other things. This is Mach's Principle, named by Einstein after the Viennese physicist Ernst Mach. There is no reference to space or time as things themselves.

metric   The formula for determining the distance between two points is the metric. It is characteristic of the shape of the surface or space in which the points are located. For example, on a flat two-dimensional surface, the metric is simply the Pythagorean theorem.

modus tollens   This is the name of a particular form of deductive argument that is always valid. Being valid means that the truth of the premises guarantees that the conclusion is also true. The specific form of *modus tollens* is as follows.

premise	If P then Q
premise	Q is false
conclusion	P is false

orbit equation (Equation (3.7))   The speed v of an object in circular orbit around a planet or star of mass M is determined by the radius r of the orbit. The specific relation of these variables is in the orbit equation.

$$v^2 = M/r$$

Note that this does not depend on the mass of the orbiting object.

parallax   The direction to a distant object changes when the visual perspective changes. This is parallax. Stellar parallax refers to the change in angular location of a particular star that results from the Earth moving.

perihelion   Planets follow elliptical orbits around the Sun, and so their distance from the Sun varies around the orbit. The point at which the orbit is closest to the Sun is the perihelion.

potential energy   Potential energy is energy stored in virtue of an object's position in a force field. A stretched rubber band has potential energy because the end is pulled out against the restoring force. In this position it has the potential to do work and turn the energy into motion, kinetic energy. An elevated stone has potential energy because it is positioned out from the gravitational force; it is higher than the ground in the gravitational field.

precession   The orientation of an elliptical orbit may slowly shift such that the semi-major axis of the ellipse advances with each orbit. This is precession.

Principle of Equivalence   The effects produced by an accelerating reference frame are indistinguishable from the effects produced by gravity. This is Einstein's Principle of Equivalence. Earlier versions of the principle noted that a heavy object falls at the same rate as a light one, and that the inertial mass and gravitational mass of an object are identical.

Principle of General Covariance   The laws of physics are the same in all reference frames. This Principle of General Covariance is the generalization of the Principle of Relativity, making laws invariant in all reference frames and not just inertial frames.

Principle of Relativity   The laws of physics are the same in all inertial reference frames. A corollary to this is that no experiment within a system can detect whether the system is uniformly moving. Both are statements of the Principle of Relativity.

reference frame   A coordinate system determined by reference to some object or feature of an object is a reference frame. The reference frame is stationary with respect to the object.

rosette   The loop-de-loop path that results by drawing a circle with its center moving along the circumference of a larger circle is a rosette. The path periodically crosses itself in a way that creates petals. Planetary orbits in the Ptolemaic model are rosettes.

rotation curve   A graph of the rotational speeds of objects in a rotating system like the Solar system or a rotating galaxy as a function of their distance from the center is a rotation curve.

scalar   A scalar is a value of a property that has no direction associated with it. There is only a magnitude and appropriate units. Temperature is a scalar. So is mass.

spacetime   Events happen at a place and at a time, that is, in space and time. The four-dimensional (three for space and one for time) array of all possible locations of events is spacetime.

spacetime diagram   A graphical representation of events at their locations in space and time is a spacetime diagram. To draw this on a two-dimensional page requires showing only one of the spatial dimensions, using the second dimension to represent the time of events.

spacetime interval   This is analogous to the distance between points; it is the spacetime distance between events. Just as the distance between two points in space is the same in any coordinate system, the spacetime interval is invariant, the same value regardless of the spacetime coordinate system used.

underdetermination   When there is not enough information in the evidence to determine if a theory is true or false the evidence underdetermines the theory. Underdetermination of theory by evidence is the reason theories cannot be decisively proven or disproven.

vector   A vector is a value of a property that has both a magnitude and a direction. Velocity is a vector, since it expresses both speed (as in meters per second) and direction (for example, due east).

void point   A void point is an empty point in space that has an essential role in determining a planetary orbit. For example, the center of an epicycle in the Ptolemaic planetary model is a void point.

weight   The force of gravity on an object is its weight. Near the surface of the Earth, the weight of an object with mass m is mg, the mass times the acceleration of gravity.

worldline   The continuous sequence of points in spacetime occupied by an object is its worldline.

Bibliography

Rewarding Reading on Gravity and Scientific Method

Barbour, J. (2001). *The Discovery of Dynamics*. Oxford: Oxford University Press.
    This is a history of the merger of astronomy and physics, from Aristotle to Newton, in a clear, lively presentation with impressive attention to detail. Barbour writes "from a Machian point of view," laying the foundations for relativity without distorting the history.

Chalmers, A. (1999). *What is This Thing Called Science?*, third edition. Indianapolis: Hackett Publishing Company.
    Philosophers of science have a lot to say about the methods and limitations of science. Here is a readable guide to topics such as underdetermination, instrumentalism, testing, and interpretation of evidence.

Cushing, J. (1998). *Philosophical Concepts in Physics*. Cambridge: Cambridge University Press.
    From Aristotle to the present, this explains the fundamentals of physics, including mechanics, electrodynamics, and quantum mechanics, in historical and methodological context. There is some math, and this makes the book a good source for understanding the details of derivation of important results.

Gates, E. (2009). *Einstein's Telescope*. New York: W.W. Norton.
    Gravitational lensing and dark matter are explained with enthusiasm and insight by a respected astronomer. Here is where you can get more details on the bullet cluster and the evidence for the existence of dark matter, as well as the different ideas of just what dark matter is.

Kuhn, T. (1957). *The Copernican Revolution*. Cambridge, MA: Harvard University Press.
    Kuhn is most famous for *The Structure of Scientific Revolutions* in which he introduces and develops the concept of a paradigm, but his ideas began with this study of the Copernican revolution. The history, from Aristotle to Kepler – and a little Newton – is helpful for understanding the scientific details and provocative for thinking about science.

Schutz, B. (2003). *Gravity From the Ground Up*. Cambridge: Cambridge University Press.
    This is where to go for more detail on the foundational concepts of gravity from Newton to relativity, with little reliance on mathematics. Modern cosmology and astrophysics are developed around the influence gravity.

Weinberg, S. (1992). *Dreams of a Final Theory*. New York: Vintage Books.
    This is not explicitly about scientific method, but it has some of the clearest insights on scientific method you will find. It makes a genuine contribution to philosophy of science, despite a chapter called "Against Philosophy." Weinberg is a Nobel Prize winning physicist, and the book is about the hope of unifying all forces of nature into one theory.

Additional References

Baum, R. and W. Sheehan (1997). *In Search of Planet Vulcan*. New York: Plenum.

Bentley, R. (1838). *Works of Richard Bentley*, volume 3. London: AMS Press.

Bridgman, P. (1955). *Reflections of a Physicist*. New York: Philosophical Library.

Clark, R. (1972). *Einstein: The Life and Times*. New York: Avon Books.

Clowe, D., Bradac, M., Gonzalez, A., *et al.* (2006). A direct empirical proof of the existence of dark matter. *The Astrophysical Journal*, **648**, L109–L113.

Copernicus, N. (1952, first published 1543). *On the Revolutions of the Heavenly Spheres*, translated by C. Wallis. In *Great Books of the Western World*, volume 16, ed. R. Hutchins: Chicago, London, Toronto: Encyclopedia Britannica, pp. 497–838.

Dear, P. (2009). *Revolutionizing the Sciences*, second edition. Princeton: Princeton University Press.

Einstein, A. (1920). *Relativity: the Special and General Theories*. New York: Henry Holt.

Einstein, A. (1936). Physics and reality. *The Journal of the Franklin Institute* **221** (3), 349–382.

Eisenstein, E. (1979). *The Printing Press as an Agent of Change*. New York: Cambridge University Press.

Feyerabend, P. (1993). *Against Method*, third edition. New York: Verso.

Galileo (1953, originally published in 1632). *Dialogue Concerning the Two Chief World Systems*, translated by Stillman Drake. Berkeley, CA: University of California Press.

Hoskin, M. ed. (1999). *The Cambridge Concise History of Astronomy*. Cambridge: Cambridge University Press.

Hoskin, M., Gingerich, O. (1999). Medieval Latin astronomy. In *The Cambridge Concise History of Astronomy*, ed. M. Hoskin. Cambridge: Cambridge University Press, pp. 68–93.

Jeffreys, H. (1919). On the crucial test of Einstein's Theory of Gravitation. *Monthly Notices of the Royal Astronomical Society*, **80** (2), 138–154.

Kepler, J. (1952, first published 1618). *Epitome of Copernican Astronomy*, translated by C. Wallis. In *Great Books of the Western World, volume 16*, ed. R. Hutchins. Chicago, London, Toronto: Encyclopedia Britannica, pp. 839–1004.

Kepler, J. (1992, first published 1609). *New Astronomy*, translated by W. Donahue. Cambridge: Cambridge University Press,

Kuhn, T. (1977). *The Essential Tension*. Chicago: University of Chicago Press.

Kuhn, T. (1996). *The Structure of Scientific Revolutions*, third edition. Chicago: University of Chicago Press.

Lange, M. (2002). *An Introduction to the Philosophy of Physics: Fields, Energy, and Mass*. Oxford: Blackwell.

Lloyd, G. (1974). *Early Greek Science: Thales to Aristotle*. New York: Norton.

Mach, E. (1911). *History and Root of the Principle of the Conservation of Energy*. Chicago: The Open Court Publishing.

Moore, P. (1996). *The Planet Neptune*, second edition. New York: Halstead Press.

Newton, I. (1995, first published 1687). *The Principia*, translated by A. Motte. Amherst, NY: Prometheus Books.

Ohanian, H. (2008). *Einstein's Mistakes: The Human Failings of Genius*. New York: Norton.

O'Neill, I. (2014). Big Bang, Inflation, Gravitational Waves: What It Means. http://news.discovery.com/space/astronomy/big-bang-inflation-gravitational-waves-what-it-all-means-140317.htm

Pais, A. (1982). *Subtle Is the Lord*. Oxford: Clarendon Press.

Reichenbach, H. (1958). *The Philosophy of Space & Time*, translated by M. Reichenbach and J. Freund. New York: Dover Publications.

Roseveare, N. T. (1982). *Mercury's Perihelion, from Le Verrier to Einstein*. Oxford: Oxford University Press.

Sachs, M. (1973). *The Field Concept in Contemporary Science*. Springfield: Charles C. Thomas.

Schwarzschild, B. (2006). Collision between galaxy clusters unveils striking evidence of dark matter. *Physics Today*, **59** (11), 21–24.

Sciama, D. (1969). *The Physical Foundations of General Relativity*. New York: Doubleday.

Sklar, L. (1974). *Space, Time, and Spacetime*. Berkeley, CA: University of California Press.

Standage, T. (2000). *The Neptune File*. New York: Walker & Company.

Turnbull, H., Scott, J., Hall, A. (1959). *The Correspondence of Isaac Newton*, volume 1. Cambridge: Cambridge University Press.

van den Bergh, S. (1999). The early history of dark matter. *Publications of the Astronomical Society of the Pacific*, **111** (760), 657–660.

Vitruvius (2001, original *c*. first century BC). *Ten Books on Architecture*, translated by I. Rowland, edited by I. Rowland and T. Howe. Cambridge: Cambridge University Press.

Will, C. (1986). *Was Einstein Right?* New York: Basic Books.

Index